GRADE **3**

Standardized Test Tutor

MATH

Practice Tests With Problem-by-Problem Strategies and Tips That Help Students Build Test-Taking Skills and Boost Their Scores

Michael Priestley

Editor: Mela Ottaiano
Cover design: Brian LaRossa
Interior design: Creative Pages, Inc.
Interior illustrations: Creative Pages, Inc.

ISBN-13: 978-0-545-09605-8
ISBN-10: 0-545-09605-7
Copyright © 2009 by Michael Priestley
All rights reserved. Published by Scholastic Inc.
Printed in the U.S.A.

2   3   4   5   6   7   8   9   10   40   15 14 13 12 11 10

Contents

Welcome to *Test Tutor*!

Students in schools today take a lot of tests, especially in reading and math. Some students naturally perform well on tests, and some do not. But just about everyone can get better at taking tests by learning more about what's on the test and how to answer the questions. How many students do you know who could benefit from working with a tutor? How many would love to have someone sit beside them and help them work their way through the tests they have to take?

That's where *Test Tutor* comes in. The main purpose of *Test Tutor* is to help students learn what they need to know in order to do better on tests. Along the way, *Test Tutor* will help students feel more confident as they come to understand the content and learn some of the secrets of success for multiple-choice tests.

The Test Tutor series includes books for reading and math in a range of grade levels. Each *Test Tutor* book in mathematics has three full-length practice tests designed specifically to resemble the state tests that students take each year. The math skills measured on these practice tests have been selected from an analysis of the skills tested in ten major states, and the questions have been written to match the multiple-choice format used in most states.

The most important feature of this book is the friendly Test Tutor. He will help students work through the tests and achieve the kind of success they are looking for. This program is designed so students may work through the tests independently by reading the Test Tutor's helpful hints. Or you may work with the student as a tutor yourself, helping him or her understand each problem and test-taking strategy along the way. You can do this most effectively by following Test Tutor's guidelines included in the pages of this book.

Three Different Tests

There are three practice tests in this book: Test 1, Test 2, and Test 3. Each test has 42 multiple-choice items with four answer choices (A, B, C, D). All three tests measure the same skills in almost the same order, but they provide different levels of tutoring help.

Test 1 provides step-by-step guidance to help students work through each problem, as in the sample on the next page. The tips in Test 1 are detailed and thorough, and they are written specifically for each math item to help students figure out how to solve the problem.

Sample 1

Last year, 1,250 people lived in Marston, and 970 people lived in Alburg. How many more people lived in Marston than in Alburg?

Ⓐ 180

Ⓑ 280

Ⓒ 320

Ⓓ 2,220

To find the difference, subtract the number of people in Alburg from the number of people in Marston.

Test 2 provides a test-taking tip for each item, as in the sample below, but the tips are less detailed than in Test 1. They help guide the student toward the solution to each problem without giving away too much. Students must take a little more initiative.

Sample 2

Max wants to buy a skateboard that costs $62.50. He has saved $59.00 so far. How much more money does he need to buy the skateboard?

Ⓐ $3.50

Ⓑ $4.50

Ⓒ $17.50

Ⓓ $121.50

Look for key words to help you understand the question. In this question, the key words are *How much more.*

Test 3 does not provide test-taking tips. It assesses the progress students have made. After working through Tests 1 and 2 with the help of the Test Tutor, students should be more than ready to score well on Test 3 without too much assistance. Success on this test will help students feel confident and ready for taking real tests.

Other Helpful Features

In addition to the tests, this book provides some other helpful features. First, on page 57, you will find an **answer sheet**. When students take the tests, they may mark their answers by filling in bubbles on the test pages. Or, they may mark their answers on a copy of the answer sheet instead, as they will be required to do in most standardized tests. You may want to have students mark their answers on the test pages for Test 1 and then use an answer sheet for Tests 2 and 3 to help the student get used to filling in bubbles.

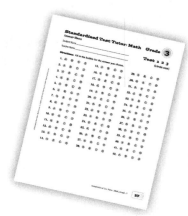

Second, beginning on page 59, you will find a detailed **answer key** for each test. The answer key lists the correct response and explains how to solve the problem. It also identifies the skill tested by each question, as in the sample below.

Answer Key for Sample 1

Correct response: **B**
(Add and subtract whole numbers)
   To find the difference in populations, subtract the number of people in Alburg (970) from the number of people in Marston (1,250): $1,250 - 970 = 280$.

Incorrect choices:

A reflects an error in subtracting $1,250 - 970$ (borrowing "2" from "12" instead of "1").

C is the result of an error in subtracting $1,250 - 970$ as the student subtracts "5" from "7" and "9" from "12."

D is the result of adding $1,250 + 970$ instead of subtracting.

As the sample indicates, question 1 measures the student's ability to add and subtract whole numbers. This information can help you determine which skills the student has mastered and which ones still cause difficulty.

Finally, the answer key explains why each incorrect answer choice, or "distractor," is incorrect. This explanation can help reveal what error students might have made. For example, one distractor in an addition problem might be the result of subtracting two numbers instead of adding them together. Knowing this could help the student understand that he or she used the wrong operation.

At the back of this book, you will find two scoring charts. The **Student Scoring Chart** can help you keep track of each student's scores on all three tests and in different subtests, such as "Number and Number Sense" or "Measurement and Geometry." The **Classroom Scoring Chart** can be used to record the scores for all students on all three tests, illustrating the progress they have made from Test 1 to Test 3. Ideally, students should score higher on each test as they go through them. However, keep in mind that students get a lot of tutoring help on Test 1, some help on Test 2, and no help on Test 3. So, if a student's scores on all three tests are all fairly similar, that could still be a very positive sign that the student is better able to solve problems independently and will achieve even greater success on future tests.

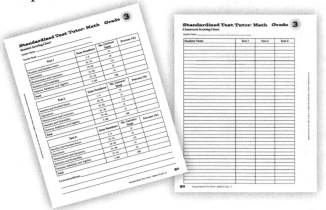

Test Tutor says:

Directions: Read each question. Look at the Test Tutor's tip for answering the question. Then find the answer. You may do your work on this page or on scrap paper. Mark your answer by filling in the bubble.

1. Which of these is the same as 2,304?

Ⓐ two thousand three hundred forty

Ⓑ two thousand three hundred four

Ⓒ two hundred thirty-four

Ⓓ twenty thousand three hundred four

> Read the number aloud. Then find the words that are the same as what you read.

2. What is six thousand eighty-four in standard form?

Ⓐ 684

Ⓑ 6,840

Ⓒ 6,804

Ⓓ 6,084

> To write this number in standard form, write each part of the number first. Then put the parts together. For example, one thousand has 3 zeros, so "six thousand" is 6,000.

3. Which set of numbers is in order from least to greatest?

Ⓐ 641, 593, 657, 579

Ⓑ 657, 641, 593, 579

Ⓒ 579, 593, 641, 657

Ⓓ 593, 579, 641, 657

> Look at the first number in each answer choice. When ordering numbers from *least to greatest*, the smallest number should come first.

4. Laura is counting by 2's. What number should come next after "twenty-four"?

Ⓐ twenty-two

Ⓑ twenty-five

Ⓒ twenty-six

Ⓓ twenty-eight

> Count by 2's aloud, starting with 20, and write the numbers as you count.

Standardized Test Tutor: Math (Grade 3) © 2009 by Michael Priestley, Scholastic Teaching Resources

Test
Tutor
says:

5. Marcus uses these blocks to show how many third graders go to his school.

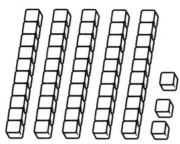

What number is shown?

Ⓐ 51

Ⓑ 53

Ⓒ 531

Ⓓ 503

Remember to look at each group of blocks. Each long block group equals 10 and each short block equals 1.

6. Which point on the number line is at $\frac{1}{2}$?

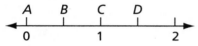

Ⓐ point *A*

Ⓑ point *B*

Ⓒ point *C*

Ⓓ point *D*

To read the number line correctly, count the number of hash marks between 0 and 1 to figure out what each mark represents.

7. Mike eats $\frac{1}{3}$ of a pizza, as shown.

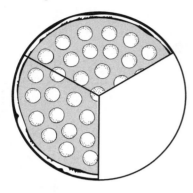

In each figure below, the shaded part represents a fraction. Which fraction is equal to the amount of pizza Mike ate?

(A)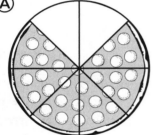

(C)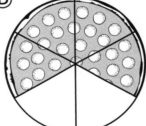

(B)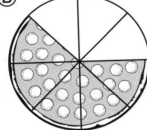

(D)

Standardized Test Tutor: Math (Grade 3) © 2009 by Michael Priestley, Scholastic Teaching Resources

Standardized Test Tutor: Math (Grade 3) © 2009 by Michael Priestley, Scholastic Teaching Resources

8. In 6,501, what does the 5 represent?

Ⓐ 5

Ⓑ 50

Ⓒ 500

Ⓓ 5,000

> To find the place value of one digit in a number, write out the number in expanded form. For example, the 6 in this number represents 6,000.

9. Gina has $\frac{3}{4}$ of a dollar. How much money does she have?

Ⓐ $0.25

Ⓑ $0.34

Ⓒ $0.43

Ⓓ $0.75

> Keep in mind that there are four quarters in one dollar, and each quarter is worth 25 cents.

10. A movie theater sold 2,384 adult tickets and 1,557 child tickets. How many tickets did the theater sell all together?

Ⓐ 827

Ⓑ 3,841

Ⓒ 3,941

Ⓓ 4,941

> The words *all together* tell you to find the total number of tickets sold. Add the number of adult tickets and the number of child tickets.

11. The town of Smithburg has 1,238 people. The town of Clarksville has 893 people. How many more people live in Smithburg than in Clarksville?

Ⓐ 335

Ⓑ 345

Ⓒ 445

Ⓓ 1,345

> Look at the words *how many more*. They tell you to find the difference, so subtract the number of people in Clarksville from the number of people in Smithburg.

12. Niko buys 6 boxes of crackers. Each box has 24 crackers. How many crackers does Niko have?

Ⓐ 30

Ⓑ 124

Ⓒ 132

Ⓓ 144

> To find the total number of crackers, multiply the number of boxes by the number of crackers in each box.

13. Jonelle ate $\frac{1}{8}$ of a pie. Her sister ate $\frac{2}{8}$ of the pie. How much of the pie did Jonelle and her sister eat?

Ⓐ $\frac{1}{8}$

Ⓑ $\frac{3}{8}$

Ⓒ $\frac{1}{4}$

Ⓓ $\frac{3}{16}$

> Notice *and* in the third sentence. It tells you to add. Add the two fractions.

14. What is 8,579 rounded to the nearest hundred?

Ⓐ 8,570

Ⓑ 8,580

Ⓒ 8,500

Ⓓ 8,600

> Remember that a number that is 5 or more is rounded upward and a number less than 5 is rounded downward.

15. A garden has 61 blue flowers and 118 white flowers. Which is the best estimate of how many flowers are in the garden?

Ⓐ 160

Ⓑ 170

Ⓒ 180

Ⓓ 190

> To estimate the answer, first round each number to the nearest ten and then add.

Standardized Test Tutor: Math (Grade 3) © 2009 by Michael Priestley, Scholastic Teaching Resources

Standardized Test Tutor: Math (Grade 3) © 2009 by Michael Priestley, Scholastic Teaching Resources

16. Chandra writes the equation $17 \times 4 = 68$. What is another way to write this equation?

Ⓐ $17 + 17 + 17 + 17 = 68$

Ⓑ $17 \times 17 \times 17 \times 17 = 68$

Ⓒ $17 + 17 + 4 + 4 = 68$

Ⓓ $17 \div 4 = 68$

> Remember that multiplication is repeated addition.

17. Mr. Montoya drove 54 miles each day for 5 days. On the next day, he drove 231 miles. How many miles did Mr. Montoya drive all together?

Ⓐ   270

Ⓑ   290

Ⓒ   481

Ⓓ   501

> Find the number of miles for the first five days by multiplying. Then take that number and add the miles from the next day.

18. James buys 4 bags of carrots. Each bag has 16 carrots. He uses 18 carrots to make soup. How many carrots does he have left?

Ⓐ   2

Ⓑ   38

Ⓒ   46

Ⓓ   82

> First, find out the total number of carrots. Then subtract. Write a number sentence. Then see if your answer is one of the choices.

19. Which clock shows the same time as the clock below?

Say each answer choice to yourself as you read it to find the time that matches the time on the clock face.

Ⓐ

Ⓒ

Ⓑ

Ⓓ

20. One foot equals 12 inches. A rope is 24 inches long. How long is the rope in feet?

Ⓐ   2 feet

Ⓑ   12 feet

Ⓒ   24 feet

Ⓓ   288 feet

To find the number of feet, divide the number of inches the rope measures by 12.

21. Which unit could best be used to measure the length of a book?

Ⓐ   pounds

Ⓑ   centimeters

Ⓒ   liters

Ⓓ   kilometers

Think about which unit of measurement is used to measure length.

Standardized Test Tutor: Math (Grade 3) © 2009 by Michael Priestley, Scholastic Teaching Resources

22. What is the length of the pencil?

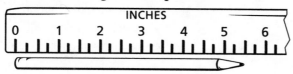

Ⓐ 5 inches

Ⓑ 5.25 inches

Ⓒ 5.5 inches

Ⓓ 6 inches

23. A rectangular field is 160 feet wide and 360 feet long.

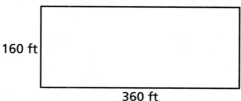

160 ft

360 ft

What is the perimeter of the field?

Ⓐ 520 feet

Ⓑ 680 feet

Ⓒ 880 feet

Ⓓ 1,040 feet

24. Tyrone has 10 quarters and 2 dimes. How much money does he have?

Ⓐ $1.20

Ⓑ $2.20

Ⓒ $2.50

Ⓓ $2.70

Standardized Test Tutor: Math (Grade 3) © 2009 by Michael Priestley, Scholastic Teaching Resources

Test Tutor says:

25. How many right angles are in an equilateral triangle?

- Ⓐ 0
- Ⓑ 1
- Ⓒ 2
- Ⓓ 3

> Remember that an equilateral triangle has three equal sides and three equal angles.

26. Which figure shows a line of symmetry?

Ⓐ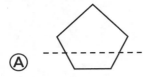

Ⓑ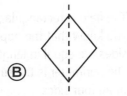

Ⓒ

Ⓓ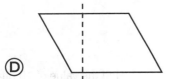

> Look for a line that divides the figure exactly in half.

27. How many faces does a cube have?

- Ⓐ 3
- Ⓑ 4
- Ⓒ 6
- Ⓓ 8

> Sketch a picture of a cube to help you figure out the number of sides.

Standardized Test Tutor: Math (Grade 3) © 2009 by Michael Priestley, Scholastic Teaching Resources

28. Which point on the grid is at (2, 5)?

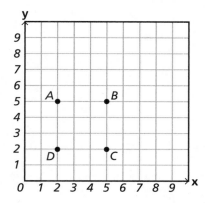

  Ⓐ point *A*   Ⓒ point *C*

  Ⓑ point *B*   Ⓓ point *D*

29. Franz buys 8 quarts of ice cream. How many cups of ice cream does he have?

  Ⓐ 2

  Ⓑ 4

  Ⓒ 16

  Ⓓ 32

30. Which two shapes can be used to make the figure below?

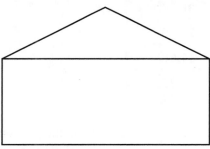

  Ⓐ rectangle and right triangle

  Ⓑ rectangle and isosceles triangle

  Ⓒ two pentagons

  Ⓓ pentagon and isosceles triangle

31. The bar graph shows the favorite sports of the students in Mrs. Li's class.

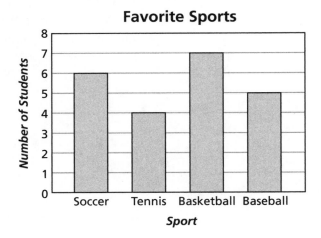

Favorite Sports

How many more students chose soccer than chose baseball?

Ⓐ 1

Ⓑ 2

Ⓒ 5

Ⓓ 6

32. Three stores charge different prices for drinks, as shown in the chart.

	Bob's Store	**Quick Stop**	**Trudy's**
Soda	$1.10	$1.40	$0.90
Milk	$1.80	$1.30	$1.60
Juice	$1.00	$1.70	$1.25

Which drink at which store has the lowest price?

Ⓐ Juice at Bob's Store

Ⓑ Soda at Trudy's

Ⓒ Milk at Quick Stop

Ⓓ Soda at Bob's Store

Standardized Test Tutor: Math (Grade 3) © 2009 by Michael Priestley, Scholastic Teaching Resources

33. Vida spins the arrow on the spinner 20 times.

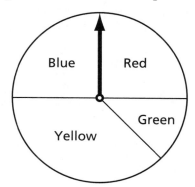

She marks the results in a tally chart.

Spinner Results	
Red	√ √ √ √ √
Yellow	√ √ √ √ √ √ √
Green	√ √ √
Blue	√ √ √ √ √

Vida spins the arrow one more time. Based on her data, it is equally likely that the arrow will land on —

(A) red or yellow

(B) blue or yellow

(C) blue or red

(D) yellow or green

Find the two colors on the spinner that the arrow landed on the same number of times.

34. Raja has a bag of 20 candies. There are 10 red candies and 10 green candies. Raja picks one candy without looking. What is the probability that she picks a red candy?

(A) $\frac{1}{2}$

(B) $\frac{1}{5}$

(C) $\frac{1}{10}$

(D) $\frac{1}{20}$

Note that half of the candies in the bag are red and half are green.

35. Mrs. Goss puts ten identical boxes on her desk. Seven boxes are empty, two have erasers, and one has stickers. Peter chooses a box. Which word best describes the chance that he chooses an empty box?

Ⓐ   certain

Ⓑ   likely

Ⓒ   unlikely

Ⓓ   impossible

> Look back at how many of the ten boxes are empty.

36. Dana takes a spelling test each week. The table shows her scores for four weeks.

Dana's Test Scores	
Week	**Score**
1	61
2	68
3	74
4	79
5	?

Based on the data, which is a good prediction about Dana's score in Week 5?

Ⓐ   It will be higher than 79.

Ⓑ   It will be lower than 79.

Ⓒ   It will be equal to 79.

Ⓓ   It will be equal to 74.

> Look for a trend in the data. Are the scores going upward, going downward, or staying the same?

Standardized Test Tutor: Math (Grade 3) © 2009 by Michael Priestley, Scholastic Teaching Resources

37. Every year, Curtis's allowance increases by $2.00. Which table shows this pattern?

Ⓐ

Year	1	2	3
Allowance	$2.00	$4.00	$8.00

Ⓑ

Year	1	2	3
Allowance	$3.00	$6.00	$12.00

Ⓒ

Year	1	2	3
Allowance	$2.00	$2.00	$2.00

Ⓓ

Year	1	2	3
Allowance	$4.00	$6.00	$8.00

> Compare columns 1, 2, and 3 in each table.

38. Look at the number pattern below.

2, 8, 14, 20, 26, . . .

How can you find the next number in the pattern?

Ⓐ   Add 6 to 26.

Ⓑ   Subtract 6 from 26.

Ⓒ   Multiply 26 by 2.

Ⓓ   Add all of the numbers together.

> Figure out what you have to do to the first number in the pattern to get the second number. Then try it on some of the other numbers in the pattern to see if it works.

39. What number makes this number sentence true?

$$6 + 4 = \underline{} \times 2$$

Ⓐ   2

Ⓑ   3

Ⓒ   4

Ⓓ   5

> Look for the number that will make the two sides of the number sentence equal.

40. Raul has 12 playing cards in his hand. After he puts down some of his cards, he has 8 cards left in his hand. If C = the number of cards Raul put down, which number sentence shows this event?

Ⓐ $12 + 8 = C$

Ⓑ $12 + C = 8$

Ⓒ $12 - C = 8$

Ⓓ $12 \div C = 8$

Write a number sentence to represent the situation. Then look at the answer choices to find your number sentence.

41. Apples cost $1.00 each. Which graph shows the cost of buying apples?

Keep in mind that the number of apples and the cost of the apples should increase at the same rate.

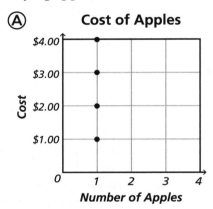

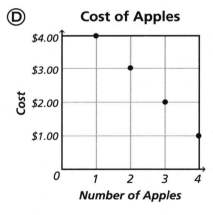

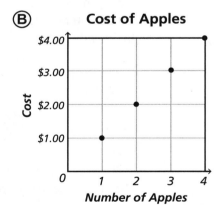

Test Tutor says:

42. Jefferson School has 912 students. The principal wants to divide the students into 28 equal groups. Which is the best estimate of how many students will be in each group?

(A) 300

(B) 60

(C) 30

(D) 4

To find an estimate, round each number and then divide.

End of Test 1 **STOP**

Directions: Read each question. Look at the Test Tutor's tip for answering the question. Then find the answer. You may do your work on this page or on scrap paper. Mark your answer by filling in the correct bubble.

1. Which of these is the same as 4,607?

   Ⓐ four hundred sixty-seven

   Ⓑ four thousand six hundred seventy

   Ⓒ four thousand six hundred seven

   Ⓓ four thousand sixty-seven

> Read each answer choice to yourself to see what number it represents.

2. What is nine thousand fifty-three in standard form?

   Ⓐ 953

   Ⓑ 9,530

   Ⓒ 9,503

   Ⓓ 9,053

> To write this number in standard form, write each part of the number first. Then put the parts together.

3. Which set of numbers is in order from least to greatest?

   Ⓐ 514, 428, 453, 592

   Ⓑ 428, 453, 514, 592

   Ⓒ 592, 514, 453, 428

   Ⓓ 453, 428, 514, 592

> "From least to greatest" means the smallest number should come first.

4. Raffy is counting by 5's. What number should come just before "thirty"?

   Ⓐ twenty

   Ⓑ twenty-five

   Ⓒ thirty-five

   Ⓓ twenty-nine

> Count by 5's to yourself.

Standardized Test Tutor: Math (Grade 3) © 2009 by Michael Priestley, Scholastic Teaching Resources

Test Tutor says:

5. Marta has a collection of marbles. She puts some of the marbles in bags, as shown.

How many marbles does Marta have?

(A) 17

(B) 20

(C) 27

(D) 207

Remember that the bags of marbles are 10's.

6. Which figure shows $\frac{3}{8}$ shaded?

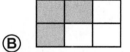

(A)

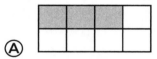

(B)

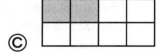

(C)

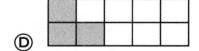

(D)

Look at all of the answer choices before you make a decision.

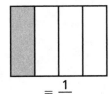

7. Look at the figures below. The shaded part of each figure represents a fraction.

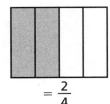 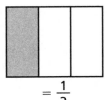

$= \dfrac{1}{4}$ $= \dfrac{2}{4}$ $= \dfrac{1}{3}$

Which set of fractions is in order from least to greatest?

Ⓐ $\dfrac{1}{4}, \dfrac{2}{4}, \dfrac{1}{3}$

Ⓑ $\dfrac{2}{4}, \dfrac{1}{3}, \dfrac{1}{4}$

Ⓒ $\dfrac{1}{3}, \dfrac{1}{4}, \dfrac{2}{4}$

Ⓓ $\dfrac{1}{4}, \dfrac{1}{3}, \dfrac{2}{4}$

> Use the figures to help compare the fractions.

8. Which number has a 4 in the hundreds place?

Ⓐ 3,428

Ⓑ 3,284

Ⓒ 2,348

Ⓓ 4,328

> Write each number in expanded form to show the value of each digit.

9. What part of the figure is shaded?

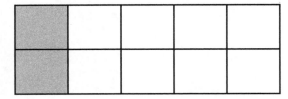

Ⓐ 0.10

Ⓑ 0.20

Ⓒ 0.02

Ⓓ 0.50

> Count the shaded boxes and the total number of boxes. Don't forget to use a decimal.

10. A factory made 4,858 magnets on Monday and 3,924 magnets on Tuesday. How many magnets did the factory make all together on those two days?

Ⓐ 7,772

Ⓑ 7,782

Ⓒ 8,782

Ⓓ 8,882

> Look for the key words *all together* to help you choose the correct operation.

11. The Gonzalez family drove 451 miles on Saturday and 374 miles on Sunday. How much farther did they drive on Saturday than on Sunday?

Ⓐ 77 miles

Ⓑ 87 miles

Ⓒ 177 miles

Ⓓ 825 miles

> Look for the key words *how much farther.*

12. Mr. Cho has 128 stickers. He divides the stickers evenly among 8 students. How many stickers does each student get?

Ⓐ 136

Ⓑ 120

Ⓒ 16

Ⓓ 14

> Remember to look for a key word to help you choose the correct operation to solve this problem.

13. Erik has $\frac{3}{4}$ cup of water. He pours out $\frac{1}{4}$ cup of water. How much water does he have left?

Ⓐ $\frac{1}{4}$ cup

Ⓑ $\frac{1}{2}$ cup

Ⓒ $\frac{3}{4}$ cup

Ⓓ 1 cup

> Write a number sentence to find the difference.

14. What is 588 rounded to the nearest ten?

Ⓐ 580

Ⓑ 590

Ⓒ 599

Ⓓ 600

It helps to look for the answer choice that is closest in value to 588.

15. A jar has 496 red jellybeans and 348 green jellybeans. Which is the best estimate of the total number of jellybeans in the jar?

Ⓐ 700

Ⓑ 750

Ⓒ 800

Ⓓ 850

Rounding to the nearest ten makes it easy to estimate the answer.

16. Denzel did this multiplication problem.

$$15 \times 6 = 90$$

What equation could he solve to check his answer?

Ⓐ $15 \div 6 = \underline{\quad}$

Ⓑ $90 - 15 = \underline{\quad}$

Ⓒ $15 + 6 = \underline{\quad}$

Ⓓ $90 \div 6 = \underline{\quad}$

If you're not sure of the answer, solve each number sentence.

17. For a field trip, 200 students were divided equally onto 4 buses. Each bus also had 5 adults. How many people were on each bus?

Ⓐ 10

Ⓑ 45

Ⓒ 55

Ⓓ 90

Remember to add the number of adults on each bus.

Standardized Test Tutor: Math (Grade 3) © 2009 by Michael Priestley, Scholastic Teaching Resources

18. Shana buys 3 shirts for $25.00 each. She also buys 2 pairs
of pants for $45.00 each. How much money does she
spend in total?

Ⓐ $70.00

Ⓑ $75.00

Ⓒ $120.00

Ⓓ $165.00

> Write a number sentence
> to solve this problem.

19. How much time has passed between Time 1 and Time 2?

Time 1   Time 2

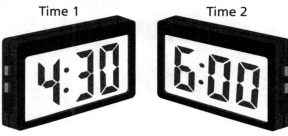

Ⓐ 1 hour 30 minutes

Ⓑ 1 hour 50 minutes

Ⓒ 2 hours 30 minutes

Ⓓ 2 hours 50 minutes

> Start by subtracting in
> hours, and then figure out
> how many minutes are left.

20. One meter equals 100 centimeters. A tree by Sal's house is
4 meters tall. How tall is the tree in centimeters?

Ⓐ 4 centimeters

Ⓑ 40 centimeters

Ⓒ 400 centimeters

Ⓓ 4,000 centimeters

> Choose the answer with
> the correct number of zeros.

21. A farmer wants to measure the weight of a pig. Which is the best unit of measurement for her to use?

Ⓐ pounds

Ⓑ liters

Ⓒ tons

Ⓓ feet

Think about the unit of measurement you use when you weigh yourself.

22. What temperature is shown on the thermometer?

Count the hash marks to read the number on the thermometer.

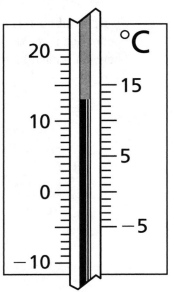

Ⓐ 15° C

Ⓑ 13° C

Ⓒ 12° C

Ⓓ 10° C

Standardized Test Tutor: Math (Grade 3) © 2009 by Michael Priestley, Scholastic Teaching Resources

23. Pat's yard is shaped like a rectangle. It is 10 meters wide and 20 meters long. What is the area of Pat's yard?

Pat's Yard

10 m

20 m

Ⓐ 30 square meters

Ⓑ 60 square meters

Ⓒ 200 square meters

Ⓓ 400 square meters

> Use the formula for finding area (width × length).

24. Four friends have some coins.

Name	Bert	Mia	Tran	Shauna
Coins	6 nickels	2 dimes	1 quarter	10 pennies

Who has the most money?

Ⓐ Bert

Ⓑ Mia

Ⓒ Tran

Ⓓ Shauna

> Find out how much each person has before you choose an answer.

25. A shape has four sides. Two sides are 5 centimeters long and two sides are 3 centimeters long. What could the shape be?

Ⓐ square

Ⓑ triangle

Ⓒ pentagon

Ⓓ rectangle

> Make a sketch of the figure to help identify it.

Standardized Test Tutor: Math (Grade 3) © 2009 by Michael Priestley, Scholastic Teaching Resources

26. Look at this picture of a house.

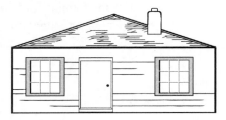

Which parts of the house are congruent shapes?

Ⓐ the door and the chimney

Ⓑ the two windows

Ⓒ the door and the windows

Ⓓ the chimney and the windows

> Remember that congruent shapes are the same size and same shape.

27. Jerry drew a shape like the one below.

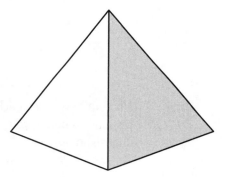

What is this shape?

Ⓐ rectangular prism

Ⓑ cube

Ⓒ sphere

Ⓓ pyramid

> Cross out any answer choices you know are wrong.

28. Look at the grid.

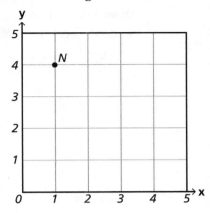

What is the location of point *N*?

Ⓐ (1, 1)

Ⓑ (1, 4)

Ⓒ (4, 1)

Ⓓ (4, 4)

> Remember that the first number in an ordered pair is the number of units across the grid.

29. A bucket weighs 1 pound. Metal balls weigh 2 pounds each. The bucket is filled with 8 metal balls. What is the weight of the bucket with the metal balls inside?

Ⓐ 9 pounds

Ⓑ 11 pounds

Ⓒ 16 pounds

Ⓓ 17 pounds

> Write a number sentence to solve this problem.

30. What figure can be made with the two shapes below?

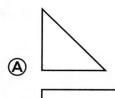

Ⓐ

Ⓑ

Ⓒ

Ⓓ

Standardized Test Tutor: Math (Grade 3) © 2009 by Michael Priestley, Scholastic Teaching Resources

31. Kara went apple picking. The graph shows how many apples of each kind she picked.

Kara's Apples

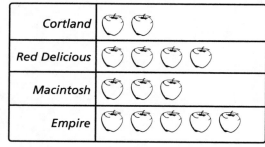

Key: ◯ = 5 apples

How many red delicious apples did Kara pick?

Ⓐ 4

Ⓒ 20

Ⓑ 15

Ⓓ 25

32. Four teams are playing a trivia game. The tally chart shows how many points each team got in two rounds of the game.

Trivia Game Score		
Team	Round A Points	Round B Points
1	√ √ √ √ √ √	√ √ √
2	√ √ √ √	√ √ √ √ √ √
3	√ √ √	√ √ √ √ √ √ √
4	√ √ √ √ √	√ √ √ √ √ √

After the two rounds, which team had the largest total score?

(A) Team 1

(B) Team 2

(C) Team 3

(D) Team 4

> Pay attention to the directions as you add points in the chart.

33. Mr. Brown has recorded the weather on his birthday for 70 years. It has been sunny 58 times and cloudy 12 times. Which is the best prediction Mr. Brown can make about the weather on his birthday next year?

(A) It is more likely to be sunny than cloudy.

(B) It is less likely to be sunny than cloudy.

(C) It is equally likely to be sunny or cloudy.

(D) It is certain to be sunny.

> Read each prediction carefully before you choose one.

34. A bag contains 34 marbles. There are 9 red, 12 yellow, 8 green, and 5 blue marbles. Naja closes her eyes and picks one marble. What color is she most likely to pick?

(A) red

(B) yellow

(C) green

(D) blue

> Compare the number of marbles of each color.

35. There are 9 gumballs in a gumball machine. There are 5 blue gumballs and 4 red gumballs. What is the probability that a red gumball will come out of the machine next?

Ⓐ 1 out of 9

Ⓑ 4 out of 5

Ⓒ 4 out of 9

Ⓓ 5 out of 9

> There are 9 gumballs in all. How many are red?

36. Emilio keeps track of how much time he spends reading and watching television after school.

Hours Spent on After-School Activities					
	Monday	Tuesday	Wednesday	Thursday	Friday
Reading	2	2	3	2	2
Watching television	2	1	0	0	1

Which statement about these data is true?

Ⓐ Emilio spent more time reading than watching television.

Ⓑ Emilio spent more time watching television than reading.

Ⓒ Emilio watched more television at the end of the week than at the beginning.

Ⓓ Emilio read more at the beginning of the week than at the end.

> Read each statement carefully and compare it to the numbers in the table.

37. The graph below shows the population of Goodville from 2000 to 2003.

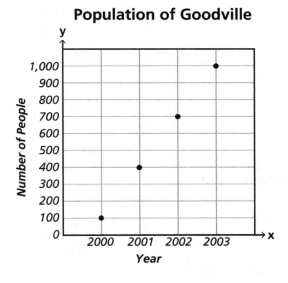

Population of Goodville

What pattern is shown in the graph?

Ⓐ The population doubles each year.

Ⓑ The population triples each year.

Ⓒ The population grows by 300 each year.

Ⓓ The population grows by 100 every 3 years.

> Look carefully at the graph to see how the numbers changed from one year to the next.

38. Look at the pattern below.

What shape comes next in the pattern?

Ⓐ

Ⓑ

Ⓒ

Ⓓ

> Figure out the pattern before you look at the answer choices.

39. What number makes this number sentence true?

$$4 \times \underline{} = 12$$

(A) 3

(B) 4

(C) 8

(D) 48

> Try each answer choice in the number sentence until you find the correct answer.

40. Debbie buys 10 boxes of crackers. Each box has the same number of crackers.

> Key:
> = the number of crackers in each box

What expression shows the number of crackers Debbie bought?

(A)  ÷ 10

(B) 10 + [ ]

(C) 10 − [ ]

(D) 10 ×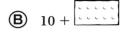

> Think about which operation represents repeating a number of items.

Standardized Test Tutor: Math (Grade 3) © 2009 by Michael Priestley, Scholastic Teaching Resources

41. Every week, Jorge's bean plant grows 3 inches. Which table could show the height of his plant over three weeks?

Look for the table that shows a change of 3 inches per week.

Ⓐ

Week	1	2	3
Plant height (inches)	3	3	3

Ⓑ

Week	1	2	3
Plant height (inches)	5	8	11

Ⓒ

Week	1	2	3
Plant height (inches)	3	6	12

Ⓓ

Year	1	2	3
Plant height (inches)	1	3	9

42. A movie theater has 18 rows of seats. Each row has 11 seats. Which is the best estimate of the total number of seats in the theater?

Use rounding to find the answer.

Ⓐ 30

Ⓑ 100

Ⓒ 200

Ⓓ 1,800

End of Test 2   **STOP**

Directions: Choose the best answer to each question. Mark your answer by filling in the correct bubble.

1. Which of these is the same as 5,098?

   Ⓐ five thousand ninety-eight

   Ⓑ five hundred ninety-eight

   Ⓒ five thousand nine hundred eight

   Ⓓ five thousand nine hundred eighty

2. What is two thousand four hundred seven in standard form?

   Ⓐ 247

   Ⓑ 2,407

   Ⓒ 2,470

   Ⓓ 2,047

3. Which set of numbers is in order from least to greatest?

   Ⓐ 715, 827, 642, 763

   Ⓑ 642, 763, 715, 827

   Ⓒ 642, 715, 763, 827

   Ⓓ 827, 763, 715, 642

4. Mina counts the letters on the sign below. What number should she say when she gets to the letter **D**?

   WELCOME TO THE GRA**N**D HOTEL

   Ⓐ fifteen

   Ⓑ sixteen

   Ⓒ eighteen

   Ⓓ seventeen

Standardized Test Tutor: Math (Grade 3) © 2009 by Michael Priestley, Scholastic Teaching Resources

5. Mr. Garcia plants 7 groups of 10 flowers and 4 single flowers. He writes $7(10) + 4(1)$ to show how many flowers he planted. How many flowers did Mr. Garcia plant?

Ⓐ 11

Ⓑ 21

Ⓒ 74

Ⓓ 704

6. Which figure shows $\frac{2}{6}$ shaded?

Ⓐ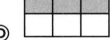

Ⓑ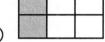

Ⓒ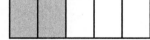

Ⓓ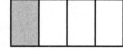

7. Look at the figures below. The shaded part of each figure represents a fraction.

$= \frac{1}{2}$　　　　$= \frac{1}{4}$　　　　$= \frac{2}{4}$

Which statement about these fractions is true?

Ⓐ $\frac{2}{4} > \frac{1}{2}$　　　　Ⓒ $\frac{1}{2} = \frac{1}{4}$

Ⓑ $\frac{1}{4} > \frac{1}{2}$　　　　Ⓓ $\frac{1}{2} = \frac{2}{4}$

8. In 4,529, what does the 2 represent?

Ⓐ 2

Ⓑ 20

Ⓒ 200

Ⓓ 2,000

9. Howard has $0.50. How much money does he have?

Ⓐ $\frac{1}{5}$ dollar

Ⓑ $\frac{1}{4}$ dollar

Ⓒ $\frac{1}{2}$ dollar

Ⓓ 5 dollars

10. On Saturday, 5,378 people visited the museum. On Sunday, 4,295 people visited the museum. How many people visited the museum on Saturday and Sunday all together?

Ⓐ 1,083

Ⓑ 9,573

Ⓒ 9,663

Ⓓ 9,673

11. A total of 8,643 people live in Starville, and 4,854 of them are children. How many are *not* children?

Ⓐ 3,789

Ⓑ 3,889

Ⓒ 4,889

Ⓓ 13,497

Standardized Test Tutor: Math (Grade 3) © 2009 by Michael Priestley, Scholastic Teaching Resources

12. A movie theater has 4 screens. There are 296 seats for each screen. How many seats does the theater have in all?

Ⓐ 74

Ⓑ 864

Ⓒ 884

Ⓓ 1,184

13. Eva added $\frac{2}{4}$ cup of milk to her cake batter. The batter was too thick, so she added another $\frac{1}{4}$ cup of milk. How much milk did she add in all?

Ⓐ $\frac{2}{4}$ cup

Ⓑ $\frac{3}{8}$ cup

Ⓒ $\frac{3}{4}$ cup

Ⓓ 1 cup

14. What is 482 rounded to the nearest ten?

Ⓐ 400

Ⓑ 480

Ⓒ 490

Ⓓ 500

15. Mrs. Liu drove 252 miles in one day. The next day she drove 149 miles. Which is the best estimate of how many miles she drove all together?

Ⓐ 300

Ⓑ 350

Ⓒ 400

Ⓓ 450

16. Mr. Goodman writes the equation $3 \times 18 =$ ___. What is another way he could write the same equation?

(A) $18 + 3 =$ ___

(B) $18 \times 3 =$ ___

(C) $18 \div 3 =$ ___

(D) $18 - 3 =$ ___

17. Risa is building 4 chairs. For each chair, she uses 16 screws for the legs and 6 screws for the back. How many screws does she use for all 4 chairs?

(A) 26

(B) 70

(C) 88

(D) 384

18. Amir makes 120 cookies. He saves 30 for his family and gives 18 to his teachers. He divides the remaining cookies evenly among his 24 classmates. How many cookies does each classmate receive?

(A) 3

(B) 5

(C) 7

(D) 48

19. Theo goes to summer camp on July 15. He comes home
on August 18.

July

Sun	Mon	Tue	Wed	Thu	Fri	Sat
						1
2	3	4	5	6	7	8
9	10	11	12	13	14	15
16	17	18	19	20	21	22
23	24	25	26	27	28	29
30	31					

August

Sun	Mon	Tue	Wed	Thu	Fri	Sat
		1	2	3	4	5
6	7	8	9	10	11	12
13	14	15	16	17	18	19
20	21	22	23	24	25	26
27	28	29	30	31		

How many weeks does Theo spend at summer camp?

Ⓐ 4

Ⓑ 5

Ⓒ 6

Ⓓ 7

20. One gallon equals 4 quarts. Mandy bought 8 gallons of juice.
How many quarts of juice did she buy?

Ⓐ 2

Ⓑ 4

Ⓒ 12

Ⓓ 32

21. Andrea lives in Melrose. Her cousin lives one hour away by car.
Which unit is best to measure the distance between Andrea's
house and her cousin's house?

Ⓐ feet

Ⓑ centimeters

Ⓒ yards

Ⓓ kilometers

22.

What is the weight of the block?

(A) $\frac{1}{2}$ pound

(B) 1 pound

(C) $1\frac{1}{2}$ pounds

(D) $2\frac{1}{2}$ pounds

23. Andy draws a map of his family's apartment and marks the length of each wall.

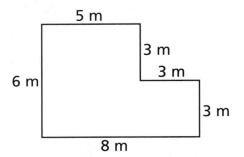

What is the perimeter of the apartment?

(A) 22 meters

(B) 25 meters

(C) 28 meters

(D) 48 meters

Standardized Test Tutor: Math (Grade 3) © 2009 by Michael Priestley, Scholastic Teaching Resources

24. Greg has some coins in his pocket, as shown below.

What is the value of the coins?

Ⓐ $0.49

Ⓑ $0.81

Ⓒ $0.89

Ⓓ $0.99

25. Which of these is a pentagon?

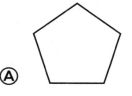

Ⓐ

Ⓑ

Ⓒ

Ⓓ

26. Which figure shows the line of symmetry on the isosceles triangle?

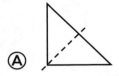

Ⓐ

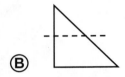

Ⓑ

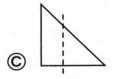

Ⓒ

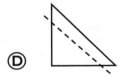

Ⓓ

27. How many faces does a square pyramid have?

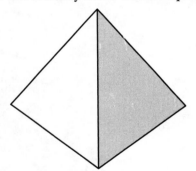

Ⓐ 3

Ⓑ 4

Ⓒ 5

Ⓓ 6

Standardized Test Tutor: Math (Grade 3) © 2009 by Michael Priestley, Scholastic Teaching Resources

28. Look at the figure on the grid.

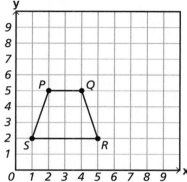

Which vertex of the figure is located at (2, 5)?

(A) P

(B) Q

(C) R

(D) S

29. Sandy buys a fish tank as shown below.

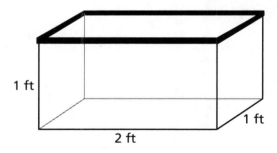

1 ft

2 ft

1 ft

What is the volume of the fish tank?

(A) 2 cubic feet

(B) 3 cubic feet

(C) 4 cubic feet

(D) 8 cubic feet

30. What shapes make up this solid object?

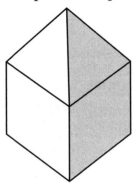

(A) square, rectangle, and triangle

(B) square and pyramid

(C) cube and triangle

(D) cube and pyramid

31. The chart shows how many games each soccer team won and lost.

Soccer Season Results		
Team	Games Won	Games Lost
Tigers	√ √ √ √	√ √ √ √ √
Bears	√ √ √ √ √ √	√ √ √
Lions	√ √ √ √ √	√ √ √ √
Eagles	√ √	√ √ √ √ √ √ √

Which teams won more games than they lost?

(A) Tigers and Bears

(B) Tigers, Bears, and Lions

(C) Bears and Lions

(D) Tigers and Eagles

Standardized Test Tutor: Math (Grade 3) © 2009 by Michael Priestley, Scholastic Teaching Resources

32. Mr. Fenton took a survey to find out what kinds of pets his students have. He found that more students own cats than own birds. Which bar graph could show his survey results?

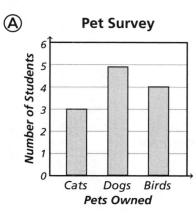

Ⓐ Pet Survey

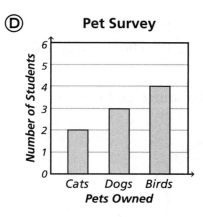

Ⓒ Pet Survey

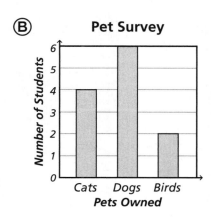

Ⓑ Pet Survey

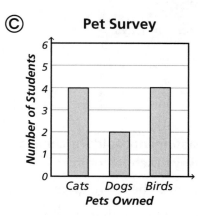

Ⓓ Pet Survey

Standardized Test Tutor: Math (Grade 3) © 2009 by Michael Priestley, Scholastic Teaching Resources

33. Dara passes a traffic light on her way to school. She keeps track of what color the light is when she reaches it each day.

Stoplight Colors	
Color	**Number of Days**
red	15
yellow	7
green	23

Based on her data, what statement could Dara make about the color of the traffic light on her way to school tomorrow?

Ⓐ The light is equally likely to be red or green.

Ⓑ The light is more likely to be red than green.

Ⓒ The light is less likely to be red than yellow.

Ⓓ The light is less likely to be yellow than green.

34. Henry has a number cube numbered 1 to 6.

If he rolls the cube once, what is the probability that it will land on 4?

Ⓐ 1 out of 6

Ⓑ 1 out of 3

Ⓒ 2 out of 6

Ⓓ 4 out of 6

Standardized Test Tutor: Math (Grade 3) © 2009 by Michael Priestley, Scholastic Teaching Resources

35. Miko has a bag of 25 colored golf balls. There are 10 red, 5 blue, 3 green, and 7 yellow golf balls in the bag. Miko closes her eyes and picks a golf ball. Which color is she *least* likely to pick?

(A) red

(B) blue

(C) green

(D) yellow

36. The chart shows the summer weather in Fayston for the past three years.

Summer Weather in Fayston			
	Sunny Days	**Cloudy Days**	**Rainy Days**
Year 1	56	24	10
Year 2	60	18	12
Year 3	54	27	9

Which is the best prediction about the weather for next summer in Fayston?

(A) There will be more cloudy days than sunny days.

(B) There will be more cloudy days than rainy days.

(C) There will be more rainy days than sunny days.

(D) There will be more rainy days than cloudy days.

37. Every person who goes to a church fair puts two raffle tickets into a jar. The table shows the relationship between numbers of people and raffle tickets in the jar.

Number of People	1	2	3	4	P
Number of Tickets	2	4	6	8	?

The letter P represents the total number of people at the fair. How could you find the total number of raffle tickets in the jar?

Ⓐ Add 2 to P.

Ⓑ Add 8 to P.

Ⓒ Multiply P by 8.

Ⓓ Multiply P by 2.

38. Look at the number pattern below.

$$2, 6, 10, 14, 18, \ldots$$

What number comes next in the pattern?

Ⓐ 19

Ⓑ 20

Ⓒ 22

Ⓓ 24

39. What number makes this number sentence true?

$$12 - 6 = \underline{} + 2$$

Ⓐ 4

Ⓑ 6

Ⓒ 8

Ⓓ 16

Standardized Test Tutor: Math (Grade 3) © 2009 by Michael Priestley, Scholastic Teaching Resources

40. Carl had 10 markers. After Sophia gave him some more markers, Carl still had fewer than 15 markers. Which inequality shows this event?

> Key:
>
> M = the number of markers Sophia gave Carl

Ⓐ $10 + 15 < M$

Ⓑ $10 + M < 15$

Ⓒ $10 + 5 < M$

Ⓓ $10 + M > 15$

41. Phil has started collecting stamps. Every month, Phil adds 2 more stamps to his collection. Which graph shows this relationship?

Ⓐ **Phil's Stamp Collection**

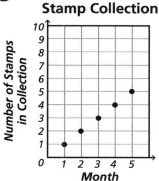

Ⓒ **Phil's Stamp Collection**

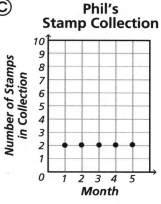

Ⓑ **Phil's Stamp Collection**

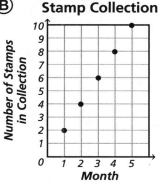

Ⓓ **Phil's Stamp Collection**

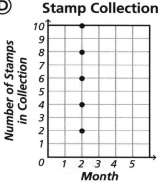

42. Quincy buys 10 bags of bagels. Each bag has the same number of bagels. What could be the total number of bagels Quincy bought?

Ⓐ 5

Ⓑ 88

Ⓒ 115

Ⓓ 150

End of Test 3   **STOP**

Standardized Test Tutor: Math (Grade 3) © 2009 by Michael Priestley, Scholastic Teaching Resources

Standardized Test Tutor: Math Grade

Answer Sheet

Student Name _____

Teacher Name _____

Test 1 2 3

(circle one)

Directions: Fill in the bubble for the answer you choose.

1. Ⓐ Ⓑ Ⓒ Ⓓ
2. Ⓐ Ⓑ Ⓒ Ⓓ
3. Ⓐ Ⓑ Ⓒ Ⓓ
4. Ⓐ Ⓑ Ⓒ Ⓓ
5. Ⓐ Ⓑ Ⓒ Ⓓ
6. Ⓐ Ⓑ Ⓒ Ⓓ
7. Ⓐ Ⓑ Ⓒ Ⓓ
8. Ⓐ Ⓑ Ⓒ Ⓓ
9. Ⓐ Ⓑ Ⓒ Ⓓ
10. Ⓐ Ⓑ Ⓒ Ⓓ
11. Ⓐ Ⓑ Ⓒ Ⓓ
12. Ⓐ Ⓑ Ⓒ Ⓓ
13. Ⓐ Ⓑ Ⓒ Ⓓ
14. Ⓐ Ⓑ Ⓒ Ⓓ

15. Ⓐ Ⓑ Ⓒ Ⓓ
16. Ⓐ Ⓑ Ⓒ Ⓓ
17. Ⓐ Ⓑ Ⓒ Ⓓ
18. Ⓐ Ⓑ Ⓒ Ⓓ
19. Ⓐ Ⓑ Ⓒ Ⓓ
20. Ⓐ Ⓑ Ⓒ Ⓓ
21. Ⓐ Ⓑ Ⓒ Ⓓ
22. Ⓐ Ⓑ Ⓒ Ⓓ
23. Ⓐ Ⓑ Ⓒ Ⓓ
24. Ⓐ Ⓑ Ⓒ Ⓓ
25. Ⓐ Ⓑ Ⓒ Ⓓ
26. Ⓐ Ⓑ Ⓒ Ⓓ
27. Ⓐ Ⓑ Ⓒ Ⓓ
28. Ⓐ Ⓑ Ⓒ Ⓓ

29. Ⓐ Ⓑ Ⓒ Ⓓ
30. Ⓐ Ⓑ Ⓒ Ⓓ
31. Ⓐ Ⓑ Ⓒ Ⓓ
32. Ⓐ Ⓑ Ⓒ Ⓓ
33. Ⓐ Ⓑ Ⓒ Ⓓ
34. Ⓐ Ⓑ Ⓒ Ⓓ
35. Ⓐ Ⓑ Ⓒ Ⓓ
36. Ⓐ Ⓑ Ⓒ Ⓓ
37. Ⓐ Ⓑ Ⓒ Ⓓ
38. Ⓐ Ⓑ Ⓒ Ⓓ
39. Ⓐ Ⓑ Ⓒ Ⓓ
40. Ⓐ Ⓑ Ⓒ Ⓓ
41. Ⓐ Ⓑ Ⓒ Ⓓ
42. Ⓐ Ⓑ Ⓒ Ⓓ

Test ① Answer Key

1. B	**10.** C	**19.** A	**28.** A	**37.** D
2. D	**11.** B	**20.** A	**29.** D	**38.** A
3. C	**12.** D	**21.** B	**30.** B	**39.** D
4. C	**13.** B	**22.** C	**31.** A	**40.** C
5. B	**14.** D	**23.** D	**32.** B	**41.** B
6. B	**15.** C	**24.** D	**33.** C	**42.** C
7. D	**16.** A	**25.** A	**34.** A	
8. C	**17.** D	**26.** B	**35.** B	
9. D	**18.** C	**27.** C	**36.** A	

Answer Key Explanations

1. Correct response: **B**
(*Read and write whole numbers*)
   The number 2,304 has "two" in the thousands place (2,000), "three" in the hundreds place (300), nothing in the tens place, and "four" in the ones place: two thousand three hundred four.

Incorrect choices:

A This number is 2,340.

C This number is 234.

D This number is 20,304.

2. Correct response: **D**
(*Read and write whole numbers*)
   The standard form of six thousand eighty-four will have a **6** in the thousands place (**6**,000), a 0 in the hundreds place, an 8 in the tens place (80), and a 4 in the ones place: 6,084.

Incorrect choices:

A This number is six hundred eighty-four.

B This number is six thousand eight hundred forty.

C This number is six thousand eight hundred four.

3. Correct response: **C**
(*Compare and order whole numbers*)
    To order these whole numbers, first consider the hundreds place, then the tens place, and then the ones place. Answer choice **C** shows the numbers in order from least to greatest.

Incorrect choices:

A has the digits in the ones place in order from least to greatest (1, 3, 7, 9).

B shows the numbers in order from greatest to least.

D reflects an error in placing 593 before 579.

4. Correct response: **C**
(*Use counting and skip counting, e.g., by 2's, 5's, 10's*)
    Since Laura is counting by 2's, the number she should say after "twenty-four" is the number that is two greater than twenty-four, or twenty-six.

Incorrect choices:

A is two less than twenty-four and would come just before it.

B would come next after twenty-four if Laura were counting by 1's.

D would come next after twenty-four if Laura were counting by 4's.

5. Correct response: **B**
(*Use grouping to represent whole numbers*)
    Five groups of 10 equal 50, and three 1's equal 3: $50 + 3 = 53$.

Incorrect choices:

A represents $50 + 1$.

C represents $500 + 30 + 1$.

D is the result of adding five 100's and three 1's.

6. Correct response: **B**
(*Identify and represent fractions*)
    Each mark on the number line represents $\frac{1}{2}$. Point *B* is located halfway between 0 and 1 on the number line, and thus is at $\frac{1}{2}$.

Incorrect choices:

A Point *A* is located at 0.

C Point *C* is located at 1.

D Point *D* is located at $1\frac{1}{2}$.

7. Correct response: **D**
(*Compare, order, convert, and find equivalent fractions*)
 To find the answer, you must find a fraction that is equivalent to $\frac{1}{3}$. The picture in choice **D** shows $\frac{2}{6}$, which can be reduced to $\frac{1}{3}$. So $\frac{1}{3}$ and $\frac{2}{6}$ are equivalent fractions.

Incorrect choices:

In **A**, the picture shows $\frac{2}{8}$, which is equivalent to $\frac{1}{4}$.

In **B**, the picture shows $\frac{3}{8}$, which is greater than $\frac{1}{3}$.

In **C**, the picture shows $\frac{1}{6}$, which is less than $\frac{1}{3}$.

8. Correct response: **C**
(*Identify and use place value*)
 This number can be written in expanded form as $6,000 + 500 + 1$. The 5 is in the hundreds place, so it represents 500.

Incorrect choices:

A would be correct if the 5 were in the ones place.

B would be correct if the 5 were in the tens place.

D would be correct if the 5 were in the thousands place.

9. Correct response: **D**
(*Read, write, and model simple decimals and relate decimal amounts of money to fractions*)
 Three-quarters of a dollar is equal to $3 \times \$0.25$, or $\$0.75$.

Incorrect responses:

A represents $\frac{1}{4}$ dollar.

B is the result of converting $\frac{3}{4}$ incorrectly to the decimal 0.34.

C is the result of converting $\frac{3}{4}$ incorrectly to the decimal 0.43.

10. Correct response: **C**
(*Add and subtract whole numbers*)
 The total number of tickets sold is the number of adult tickets plus the number of child tickets: $2,384 + 1,557 = 3,941$.

Incorrect choices:

A is the result of subtracting $2,384 - 1,557$ instead of adding.

B reflects an error in addition (not carrying the "1" to the hundreds place).

D reflects an error in addition (carrying an extra "1" to the thousands place).

11. Correct response: **B**
   (*Add and subtract whole numbers*)
   To find the difference in the two populations, subtract the number of people in Clarksville (893) from the number of people in Smithburg (1,238): $1{,}238 - 893 = 345$.

   Incorrect choices:

   A reflects an error in subtraction (in the tens column).

   C reflects an error in subtraction (in the hundreds column).

   D is the result of subtracting accurately, but retaining the 1 in the thousands place.

12. Correct response: **D**
   (*Multiply and divide whole numbers*)
   To find the total number of crackers Niko has, multiply the number of boxes of crackers times the number of crackers in each box: $6 \times 24 = 144$.

   Incorrect choices:

   A is the result of adding $6 + 24$ instead of multiplying.

   B reflects an error (not carrying the "2" from 6×4).

   C reflects an error (thinking that $6 \times 4 = 12$).

13. Correct response: **B**
   (*Add and subtract simple fractions*)
   To find the amount of pie Jonelle and her sister ate, add the fractions: $\frac{1}{8} + \frac{2}{8} = \frac{3}{8}$. When adding fractions with a common denominator, the numerators are added and the denominators stay the same.

   Incorrect choices:

   A is the amount of pie Jonelle ate $(\frac{1}{8})$.

   C is a fraction equivalent to the amount of pie Jonelle's sister ate $(\frac{2}{8} = \frac{1}{4})$.

   D is the result of adding the denominators as well as the numerators.

14. Correct response: **D**
   (*Round whole numbers to the nearest ten or hundred*)
   The number 8,579 rounded to the nearest hundred is 8,600; since 79 is greater than 50, the digit in the hundreds place (5) must be rounded up.

   Incorrect choices:

   A is the number 8,579 rounded incorrectly (downward) to the nearest ten.

   B is the result of rounding to the nearest ten instead of nearest hundred.

   C is the number 8,579 rounded incorrectly (downward) to the nearest hundred.

15. Correct response: **C**

(*Estimate using whole numbers*)

 To estimate the total number of flowers, round the number of blue flowers to 60 and the number of white flowers to 120. Then find the sum: $60 + 120 = 180$.

Incorrect choices:

A is the result of rounding the numbers incorrectly to $60 + 100$.

B is the result of rounding the numbers incorrectly to $60 + 110$.

D is the result of rounding both numbers upward (incorrectly) to $70 + 120$.

16. Correct response: **A**

(*Apply the properties of operations*)

 Multiplication can be seen as repeated addition. The equation $17 \times 4 = 68$ can be written as addition by adding four 17's: $17 + 17 + 17 + 17 = 68$.

Incorrect choices:

In **B**, the number sentence multiplies 17 by itself 4 times.

C and **D** reflect misunderstandings of the properties of operations.

17. Correct response: **D**

(*Solve multi-step problems involving whole numbers*)

 The total number of miles Mr. Montoya drove in the first 5 days is the number of miles he drove each day (54) times the number of days (5): $54 \times 5 = 270$. The number of miles he drove all together is 270 plus the number of miles he drove on the sixth day: $270 + 231 = 501$.

Incorrect choices:

A reflects the number of miles he drove in 5 days (54×5) instead of six days.

B is the result of adding $54 + 5 + 231$ (instead of multiplying 54×5).

C reflects an error in multiplying 54×5 and getting 250 instead of 270.

18. Correct response: **C**

(*Solve multi-step problems involving whole numbers*)

 The number of carrots James buys is equal to the number of bags times the number of carrots in each bag: $4 \times 16 = 64$. The number of carrots he has left after he makes the soup is the number of carrots he bought minus the number he used in the soup: $64 - 18 = 46$.

Incorrect choices:

A is the result of adding $4 + 16$ instead of multiplying, then subtracting 18.

B is the result of adding $4 + 16$ instead of multiplying, then adding 18.

D is the result of multiplying 4×16 and then adding 18 instead of subtracting.

19. Correct response: **A**

(*Tell time and find elapsed time*)

The hour hand on the clock face points just past the one, meaning that the hour is one o'clock. The minute hand points at the three, which represents a quarter of an hour, or fifteen minutes. The time shown on the clock face is 1:15.

Incorrect choices:

B mistakes the minute hand pointing at the 3 to mean 30 minutes.

C is the result of confusing the minute hand and the hour hand.

D confuses the minute hand and the hour hand and mistakes the hand pointing at the 1 to mean 10 minutes.

20. Correct response: **A**

(*Convert units of measurement, e.g., feet and inches, meters and centimeters*)

To convert the length of the rope from inches to feet, divide the number of inches by 12: 24 inches ÷ 12 inches/foot = 2 feet.

Incorrect choices:

B is the number of inches in a foot.

C is the length of the rope in inches.

D is the result of multiplying 24 × 12 instead of dividing.

21. Correct response: **B**

(*Select appropriate unit for measuring weight/mass, capacity, length, perimeter, and area*)

Centimeters are units used to measure length. Since there are about $2\frac{1}{2}$ centimeters to one inch, centimeters are appropriate for measuring the length of a book.

Incorrect choices:

A Pounds are used to measure weight, not length.

C Liters are used to measure volume or capacity, not length.

D is a measure of length, but kilometers are too large for measuring the length of a book (since a kilometer is about $\frac{6}{10}$ mile).

22. Correct response: **C**

(*Use rulers, thermometers, and other instruments to measure accurately*)

The end of the pencil is lined up with the "0" on the ruler, and the tip is lined up with "5.5," so the pencil is 5.5 inches long.

Incorrect choices:

In **A**, the measurement overlooks the 0.5 inch.

B reflects a misreading of the fractional measure as 0.25 inch instead of 0.5.

D is highest number shown on the ruler.

23. Correct response: **D**

(*Estimate and find length, perimeter, and area*)

The perimeter of a rectangle is the sum of all of its sides. This field has two sides that are 160 feet long and two sides that are 360 feet long, so its perimeter is $160 + 160 + 360 + 360 = 1,040$ feet.

Incorrect choices:

A is the result of adding only the two sides shown: $160 + 360$.

B is the result of adding two widths and one length: $160 + 160 + 360$.

C is the result of adding two lengths and one width: $160 + 360 + 360$.

24. Correct response: **D**

(*Count money, compare units of money, and solve problems involving money*)

A quarter is worth $0.25, so 10 quarters are worth $10 \times \$0.25 = \$2.50$. A dime is worth $0.10, so 2 dimes are worth $2 \times \$0.10 = \$0.20$. The total amount of money Tyrone has is $\$2.50 + \$0.20 = \$2.70$.

Incorrect choices:

A is the result of thinking that 10 quarters = one dollar and adding $1.00 + $0.20.

B is the result of thinking that 10 quarters = two dollars and adding $2.00 + $0.20.

C is the value of the 10 quarters ($10 \times \$0.25$) but does not include the value of the two dimes.

25. Correct response: **A**

(*Identify, classify, and describe plane figures and their attributes*)

An equilateral triangle has no right angles (it has three equal angles of 60 degrees).

Incorrect choices:

B A right triangle has one right angle.

C There is no kind of triangle that has 2 right angles.

D An equilateral triangle has 3 angles, but they are not right angles.

26. Correct response: **B**

(*Identify congruent figures and lines of symmetry*)

The kite shape has a line of symmetry, which divides the figure into two matching halves. This may be demonstrated by cutting out a diamond figure and folding it in half; the two halves will overlap each other perfectly.

Incorrect choices:

In **A**, the dotted line does not divide the figure into two matching halves.

In **C**, this figure does not have any lines of symmetry.

In **D**, the dotted line does not divide the figure into matching halves.

27. Correct response: **C**

(*Identify, classify, and describe solid figures and their attributes*)

A cube has six faces.

Incorrect choices:

A, **B**, and **D** represent misconceptions about cubes or a misunderstanding of the term *faces*.

28. Correct response: **A**

(*Locate points on a coordinate plane using ordered pairs*)

In an ordered pair, the first number (*x*) indicates the number of units across and the second number (*y*) indicates the number of units up or down. On the grid, point *A* is 2 units across and 5 units up.

Incorrect choices:

In **B**, point *B* is located at (5, 5).

C is the result of reading the coordinates in reverse order; point *C* is located at (5, 2).

In **D**, point *D* is located at (2, 2).

29. Correct response: **D**

(*Solve problems involving volume/capacity and weight/mass*)

One quart = 4 cups. To find the number of cups in 8 quarts of ice cream, multiply 8 quarts $\times$ 4 cups/quart = 32 cups of ice cream.

Incorrect choices:

A is the result of dividing 8 quarts by 4 cups instead of multiplying.

B is the number of cups in a quart.

C is the result of thinking that a quart equals 2 cups and multiplying 8 quarts $\times$ 2 cups.

30. Correct response: **B**

(*Recognize, compose, and decompose 2-dimensional and 3-dimensional shapes*)

This figure is made up of a rectangle on the bottom and an isosceles triangle on the top.

Incorrect choices:

A The triangle in this figure does not contain a right angle.

C This figure does not contain any 5-sided figures (pentagon).

D reflects a confusion between a pentagon and a rectangle as the shape on the bottom of the figure.

31. Correct response: **A**

(*Interpret data presented in bar graphs, pictographs, tables, tallies, and charts*)

To find the difference, subtract the number of students who chose baseball (5) from the number who chose soccer (6): $6 - 5 = 1$.

Incorrect choices:

B is the difference between the number of students who chose soccer and tennis.

C is the number of students who chose baseball.

D is the number of students who chose soccer.

32. Correct response: **B**

(*Interpret data presented in bar graphs, pictographs, tables, tallies, and charts*)

To answer this question, you must look at both the kinds of drinks and the different stores. A soda at Trudy's costs $0.90, which is the lowest price on the chart.

Incorrect choices:

A A juice at Bob's Store costs $1.00, which is higher than a soda at Trudy's.

C A milk at Quick Stop costs $1.30.

D A soda at Bob's Store costs $1.10.

33. Correct response: **C**

(*Use data to describe events as more, less, or equally likely*)

Vida's data show that the arrow landed on blue 5 times and on red 5 times; so, based on the data, these events are equally likely to happen again.

Incorrect choices:

A The spinner landed on red 5 times and yellow 7 times, so these events are not equally likely to happen again.

B The spinner landed on blue 5 times and yellow 7 times, so these events are not equally likely to happen again.

D The spinner landed on yellow 7 times and green 3 times, so these events are not equally likely to happen again.

34. Correct response: **A**

(*Describe, explain, or determine simple probabilities or outcomes of an experiment*)

There are 20 candies in all and 10 of the 20 candies are red, so the probability of choosing a red candy is $\frac{10}{20}$, or $\frac{1}{2}$.

Incorrect choices:

B reflects a random guess or a misunderstanding of how to find probability.

C is "1" over the number of red candies, which reflects a misunderstanding of probability.

D is the probability of picking either red or green.

35. Correct response: **B**

(*Describe, explain, or determine simple probabilities or outcomes of an experiment*)

Since 7 of the 10 boxes are empty and 7 out of 10 is more than one-half, it is *likely* that Peter will choose an empty box.

Incorrect choices:

A would be correct if all the boxes were empty.

C would be correct if fewer than one-half of the boxes were empty.

In **D**, the chance of picking an empty box would only be impossible if none of the boxes was empty.

36. Correct response: **A**

(*Make predictions or draw conclusions based on data*)

The data show that Dana's score has gotten higher each week (increasing by 5 to 7 points), and her last score was 79, so a good prediction for her next score is that it will be higher than 79.

Incorrect choices:

B The trend suggests that the next score will go higher, not lower.

C This is the score that Dana got in week 4.

D The trend does not support a decrease to the score from week 3.

37. Correct response: **D**

(*Identify and represent visual and number patterns using words, variables, tables, and graphs*)

The values in this table start at $4.00 and increase by $2.00 each year: $4.00 + $2.00 = $6.00; $6.00 + $2.00 = $8.00.

Incorrect choices:

A shows the allowance increasing by $4.00 between years 2 and 3.

B shows the allowance doubles each year.

C does not show an increase; the allowance stays at $2.00 each year.

38. Correct response: **A**

(*Identify, describe, and extend numerical and geometric patterns*)

Each number in the pattern is 6 greater than the number before it. To find the next number in the pattern, add 6 to the last number, 26: 26 + 6 = 32.

Incorrect choices:

B is the result of working backward in the pattern (e.g., 26 − 20 = 6) instead of forward.

C reflects an error in choosing the operation of multiplication instead of addition.

D reflects a misunderstanding of patterns.

39. Correct response: **D**

(*Solve simple number sentences or equations with one variable*)

On the left side of the number sentence, 6 + 4 is equal to 10. To make the right side of the number sentence equal to 10, you can multiply 5 × 2.

Incorrect choices:

A Multiplying by 2 would make the right side of the number sentence equal to 4.

B Multiplying by 3 would make the right side of the number sentence equal to 6.

C Multiplying by 4 would make the right side of the number sentence equal to 8.

40. Correct response: **C**

(*Use objects, symbols, and words to model concepts of variables, expressions, equations, inequalities*)

Raul started with 12 cards. Subtract the number of cards Raul put down (*C*) from the number of cards he started with (12) to find the number of cards he ended up with (8): 12 − *C* = 8.

Incorrect choices:

A does not model the problem situation; it adds the number of cards Raul started with (12) and the number of cards he has left (8).

B does not model the problem situation; it adds the number of cards he started with (12) and the number he put down (*C*).

D does not model the problem situation; it divides the number of cards he started with (12) by *C* instead of subtracting.

41. Correct response: **B**

(*Represent and describe mathematical relationships with lists, tables, charts, graphs, and diagrams*)

The cost of buying apples is the number of apples times $1.00. Buying one apple costs $1.00, buying two apples costs $2.00, and so on. The graph in answer choice **B** shows this relationship.

Incorrect choices:

In **A**, the graph shows one apple costing different amounts.

In **C**, the graph shows any number of apples costing $1.00.

In **D**, the graph shows the cost decreasing as the number of apples increases (for example, 4 apples cost $1.00 while 1 apple costs $4.00).

42. Correct response: **C**

(*Identify reasonable solutions*)

To find the best estimate, round 912 to 900 and 28 to 30, and then divide: 900 ÷ 30 = 30.

Incorrect choices:

A reflects an error in division (adding a zero or dividing 900 by 3 instead of by 30).

B reflects an error in estimation (estimating 912 as 90) and applying subtraction instead of division.

D reflects an error in computation; for example, rounding 912 to 100 and 28 to 25.

Test ② Answer Key

1. C	**10.** C	**19.** A	**28.** B	**37.** C
2. D	**11.** A	**20.** C	**29.** D	**38.** B
3. B	**12.** C	**21.** A	**30.** C	**39.** A
4. B	**13.** B	**22.** B	**31.** C	**40.** D
5. C	**14.** B	**23.** C	**32.** D	**41.** B
6. A	**15.** D	**24.** A	**33.** A	**42.** C
7. D	**16.** D	**25.** D	**34.** B	
8. A	**17.** C	**26.** B	**35.** C	
9. B	**18.** D	**27.** D	**36.** A	

Answer Key Explanations

1. Correct response: **C**
(*Read and write whole numbers*)
   The number 4,607 has "four" in the thousands place, "six" in the hundreds place, nothing in the tens place, and "seven" in the ones place: four thousand six hundred seven.

Incorrect choices:

A This number is 467.

B This number is 4,670.

D This number is 4,067.

2. Correct response: **D**
(*Read and write whole numbers*)
   The standard form of nine thousand fifty-three will have a **9** in the thousands place (**9,**000), a 0 in the hundreds place, a 5 in the tens place (5) and a 3 in the ones place: 9,053.

Incorrect choices:

A This number is nine hundred fifty-three.

B This number is nine thousand five hundred thirty.

C This number is nine thousand five hundred three.

3. Correct response: **B**

(*Compare and order whole numbers*)

To order these whole numbers, first consider the hundreds place, then the tens place, and then the ones place. Answer choice **B** shows the numbers in order from least to greatest.

Incorrect choices:

A has the digits in the tens place in order from least to greatest (1, 2, 5, 9).

C shows the numbers in order from greatest to least.

D reverses the order of the first two numbers, which should be 428 and 453.

4. Correct response: **B**

(*Use counting and skip counting, e.g., by 2's, 5's, 10's*)

Since Raffy is counting by 5's, the number before "thirty" should be the number that is 5 less than thirty, or twenty-five.

Incorrect choices:

A is ten less than thirty and would be correct if Raffy were counting by 10's.

C is the number that would come just after thirty, not before.

D is one less than thirty and would be correct if Raffy were counting by 1's.

5. Correct response: **C**

(*Use grouping to represent whole numbers*)

Two sets of 10 equal 20, and seven 1's equal 7: $20 + 7 = 27$.

Incorrect choices:

A represents one set of $10 + 7$.

B represents two sets of 10 but does not include the seven 1's.

D represents two sets of 100 and seven 1's.

6. Correct response: **A**

(*Identify and represent fractions*)

The figure in answer choice **A** is divided into 8 equal sections, and 3 of the 8 sections are shaded, representing $\frac{3}{8}$.

Incorrect choices:

In **B**, the figure shows $\frac{3}{6}$ shaded.

In **C**, the figure shows $\frac{2}{8}$ shaded.

In **D**, the figure shows $\frac{3}{10}$ shaded.

7. Correct response: **D**

(*Compare, order, convert, and find equivalent fractions*)

If two fractions have the same numerator, then the fraction with the larger denominator will be smaller. So, $\frac{1}{4} < \frac{1}{3}$, and since $\frac{2}{4} = \frac{1}{2}$, $\frac{1}{3} < \frac{2}{4}$.

Incorrect choices:

A lists $\frac{2}{4}$ before $\frac{1}{3}$ instead of after.

B lists the fractions in reverse order from greatest to least.

C lists $\frac{1}{3}$ before $\frac{1}{4}$ instead of after.

8. Correct response: **A**

(*Identify and use place value*)

The hundreds place is 3 to the left of the decimal, so the number 3,428 has a 4 in the hundreds place. It can be written in expanded form as $3,000 + 400 + 20 + 8$.

Incorrect choices:

In **B**, the number has a 2 in the hundreds place and a 4 in the ones place.

In **C**, the number has a 3 in the hundreds place and a 4 in the tens place.

In **D**, the number has a 4 in the thousands place and a 3 in the hundreds place.

9. Correct response: **B**

(*Read, write, and model simple decimals and relate decimal amounts of money to fractions*)

This figure is divided into 10 equal parts, and 2 of the 10 parts are shaded; $\frac{2}{10}$ equals 0.20.

Incorrect responses:

A would be correct if only one of the 10 parts were shaded.

C reflects an error in place value; 0.02 is $\frac{2}{100}$.

D would be correct if one-half of the parts were shaded.

10. Correct response: **C**

(*Add and subtract whole numbers*)

To find the total number of magnets made, find the sum of the number made on Monday and the number made on Tuesday: $4,858 + 3,924 = 8,782$.

Incorrect choices:

A reflects errors in addition (in the tens and the thousands columns).

B reflects an error in addition (in the thousands column).

D reflects an error in addition (in the hundreds column).

11. Correct response: **A**

   (*Add and subtract whole numbers*)

   To find the difference in the two distances, subtract the number of miles driven on Saturday (374) from the number of miles driven on Sunday (451): $451 - 374 = 77$.

   Incorrect choices:

   B reflects an error in subtraction (in the tens column).

   C reflects an error in subtraction (in the hundreds column).

   D is the result of adding $451 + 374$ instead of subtracting.

12. Correct response: **C**

   (*Multiply and divide whole numbers*)

   To find the number of stickers each student gets, divide the number of stickers (128) by the number of students (8): $128 \div 8 = 16$.

   Incorrect choices:

   A is the result of adding $128 + 8$ instead of dividing.

   B is the result of subtracting $128 - 8$ instead of dividing.

   D reflects an error in division (thinking that $48 \div 8$ is 4 instead of 6).

13. Correct response: **B**

   (*Add and subtract simple fractions*)

   To find the amount of water Erik has left, subtract the amount he poured out ($\frac{1}{4}$ cup) from the amount he started with ($\frac{3}{4}$ cup): $\frac{3}{4}$ cup $- \frac{1}{4}$ cup $= \frac{2}{4}$ cup, or $\frac{1}{2}$ cup.

   Incorrect choices:

   A is the amount of water Erik poured out.

   C is the amount of water Erik started with.

   D is the result of adding $\frac{3}{4} + \frac{1}{4}$ instead of subtracting.

14. Correct response: **B**

   (*Round whole numbers to the nearest ten or hundred*)

   The number 588 rounded to the nearest ten is 590; since 8 is greater than 5, the digit in the tens place (8) must be rounded up.

   Incorrect choices:

   A is the result of rounding down instead of up.

   C is the result of rounding each 8 to the next highest digit.

   D is the result of rounding to the nearest hundred instead of ten.

15. Correct response: **D**

(*Estimate using whole numbers*)

   To estimate the total number of jellybeans, round the number of red jellybeans to 500 and the number of green jellybeans to 350. Then find the sum: 500 + 350 = 850.

Incorrect choices:

A is the result of rounding to 400 + 300.

B is the result of rounding to 400 + 350.

C is the result of rounding to 500 + 300.

16. Correct response: **D**

(*Apply the properties of operations*)

   Multiplication and division have an inverse relationship, so if 6 × 15 = 90, then 90 ÷ 15 = 6.

Incorrect choices:

In **A**, 15 ÷ 6 is not the same as 15 × 6.

B and **C** represent misunderstandings of the properties of operations.

17. Correct response: **C**

(*Solve multi-step problems involving whole numbers*)

   To find the number of students on each bus, divide the total number of students by the number of buses: 200 ÷ 4 = 50. The total number of people on each bus is the number of students (50) plus the number of adults on each bus (5): 50 + 5 = 55.

Incorrect choices:

A is the result of dividing 200 ÷ 4 and then dividing 50 by 5 instead of adding.

B is the result of dividing 200 ÷ 4 and then subtracting (50 − 5) instead of adding.

D is the result of dividing 200 ÷ 4 and 200 ÷ 5 and then adding the results (50 + 40).

18. Correct response: **D**

(*Solve multi-step problems involving whole numbers*)

   To find the total cost of the shirts, multiply the number of shirts (3) by the cost of each shirt ($25.00). The number sentence is: 3 × $25.00 = $75.00. To find the total cost of the pants, multiply the number of pairs of pants (2) by the cost of each pair ($45.00). Set up as: 2 × $45.00 = $90.00. Shana spent a total of $75.00 + $90.00 = $165.00.

Incorrect choices:

A is the result of adding the cost of one shirt ($25.00) and one pair of pants ($45.00).

B is the cost of the shirts but does not include the cost of the pants.

C is the cost of the shirts plus the cost of only one pair of pants ($75.00 + $45.00).

19. Correct response: **A**

(*Tell time and find elapsed time*)

Since there are 60 minutes in an hour, there are 30 minutes between 4:30 and 5:00. There is one hour between 5:00 and 6:00. The total time elapsed is 1 hour 30 minutes.

Incorrect choices:

B assumes that one hour equals 100 minutes.

C adds an extra hour.

D adds an extra hour and assumes 100 minutes in an hour.

20. Correct response: **C**

(*Convert units of measurement, e.g., feet and inches, meters and centimeters*)

To convert meters to centimeters, multiply the height of the tree (4 m) times 100:
4 meters × 100 centimeters/meter = 400 centimeters.

Incorrect choices:

A is the height of the tree (4 m) changed to 4 cm.

B reflects an error in computation, by a power of ten (4 × 10).

D reflects an error in computation, by a power of ten (4 × 1,000).

21. Correct response: **A**

(*Select appropriate unit for measuring weight/mass, capacity, length, perimeter, and area*)

If you wanted to find your own weight, you would use pounds. It is also the most appropriate unit of measurement to find the weight of a pig (or a dog, cat, and so on).

Incorrect choices:

B Liters are used to measure volume or capacity, not weight.

C is a unit of weight, but it is too large for weighing a pig (one ton = 2,000 pounds).

D Feet are used to measure length, not weight.

22. Correct response: **B**

(*Use rulers, thermometers, and other instruments to measure accurately*)

Each hash mark on the thermometer represents 1°C. The mercury in the thermometer reaches to 3 hash marks above 10°C, which is 13°C.

Incorrect choices:

A, C, and **D** represent incorrect readings of the thermometer.

23. Correct response: **C**

(*Estimate and find length, perimeter, and area*)

To find the area of the rectangular yard, multiply its length times its width: 10 m × 20 m = 200 square meters.

Incorrect choices:

A is the result of adding the width and the length: 10 + 20.

B is the perimeter of the yard: (2 × 10) + (2 × 20).

D is 2 times the area of the yard.

24. Correct response: **A**

(*Count money, compare units of money, and solve problems involving money*)

To compare these four amounts, change them to decimals. Six nickels are worth 6 × $0.05 = $0.30. Two dimes are worth 2 × $0.10 = $0.20. One quarter is worth $0.25. Ten pennies are worth 10 × $0.01 = $0.10. Bert has the most money.

Incorrect choices:

B Mia has 2 dimes, or $0.20, which is less than $0.30.

C Tran has 1 quarter, or $0.25, which is less than $0.30.

D Shauna has 10 pennies, or $0.10, which is less than $0.30.

25. Correct response: **D**

(*Identify, classify, and describe plane figures and their attributes*)

A rectangle has two pairs of sides with equal lengths, so a shape that has two sides of 5 centimeters and two sides of 3 centimeters could be a rectangle.

Incorrect choices:

A A square has four equal sides.

B A triangle has three sides.

C A pentagon has five sides.

26. Correct response: **B**

(*Identify congruent figures and lines of symmetry*)

The two windows are the same shape (squares) and the same size, so they are congruent.

Incorrect choices:

A The door and the chimney are the same shape but not the same size.

C The door is a different shape from the windows.

D The chimney is a different shape from the windows.

27. Correct response: **D**
(*Identify, classify, and describe solid figures and their attributes*)
  This shape has a square base and four triangular faces, so it is a pyramid.

Incorrect choices:

A A rectangular prism has six rectangular faces.

B A cube has six sides that are squares.

C is a round shape with no sides.

28. Correct response: **B**
(*Locate points on a coordinate plane using ordered pairs*)
  In an ordered pair, the first number (x) indicates the number of units across and the second number (y) indicates the number of units up or down. On the grid, point N is 1 unit across and 4 units up.

Incorrect choices:

A This point would be located at 1 unit across and 1 unit up.

C is the result of reading the coordinates in reverse order; point N is located at (1, 4).

D This point would be located at 4 units across and 4 units up.

29. Correct response: **D**
(*Solve problems involving volume/capacity and weight/mass*)
  To find the weight of the 8 metal balls, multiply the number of balls (8) times the weight of each one (2 pounds): 8×2 pounds = 16 pounds. The weight of the bucket is 1 pound. The weight of the bucket with the balls inside is 16 pounds + 1 pound = 17 pounds.

Incorrect choices:

A is the result of calculating the weight of the balls at 1 pound each (1 pound + 8 pounds).

B is the result of adding the given numbers: 1 pound + 2 pounds + 8.

C is the weight of the 8 balls but does not include the weight of the bucket.

30. Correct response: **C**
(*Recognize, compose, and decompose 2-dimensional and 3-dimensional shapes*)
  The two shapes can be put together to form a parallelogram as shown:

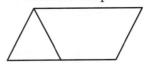

Incorrect choices:

A The two shapes can be used to form an equilateral triangle (by placing the triangle upside down at the bottom of the trapezoid) but not a right triangle.

In **B** and **D**, the figures cannot be made from the two shapes.

31. Correct response: **C**

(*Interpret data presented in bar graphs, pictographs, tables, tallies, and charts*)

There are 4 pictures of apples in the pictograph next to "Red Delicious." Each picture of an apple represents 5 apples, so Kara picked $4 \times 5 = 20$ red delicious apples.

Incorrect choices:

A is the number of pictures of apples next to "Red Delicious."

B is the number of Macintosh apples Kara picked (3×5).

D is the number of Empire apples Kara picked (5×5).

32. Correct response: **D**

(*Interpret data presented in bar graphs, pictographs, tables, tallies, and charts*)

Adding the points from Rounds 1 and 2, Team 1 had $7 + 4 = 11$ points; Team 2 had $4 + 6 = 10$ points; Team 3 had $3 + 8 = 11$ points; and Team 4 had $6 + 6 = 12$ points. Team 4 had the largest total score.

Incorrect choices:

A Team 1 had $7 + 4 = 11$ points, which is less than Team 4's score.

B Team 2 had $4 + 6 = 10$ points, which is less than Team 4's score.

C Team 3 had $3 + 8 = 11$ points, which is less than Team 4's score.

33. Correct response: **A**

(*Use data to describe events as more, less, or equally likely*)

Mr. Brown's data show that it is more often sunny on his birthday (58 times) than cloudy (12 times), so he could predict that it is more likely to be sunny than cloudy on his birthday next year.

Incorrect choices:

B The data suggest that a sunny day is more likely than a cloudy one.

C The number of sunny days and cloudy days over 70 years is not close to equal, so the likelihoods of a sunny or cloudy day are not equal.

D The data do not support any "certain" prediction.

34. Correct response: **B**

(*Describe, explain, or determine simple probabilities or outcomes of an experiment*)

In a bag of 34 marbles, there are more yellow marbles (12) than any other color of marble, so Naja is most likely to pick a yellow marble.

Incorrect choices:

A There are fewer red marbles (9) than yellow marbles (12).

C There are fewer green marbles (8) than yellow marbles (12).

D There are fewer blue marbles (5) than yellow marbles (12); blue is the least likely color to be picked.

35. Correct response: **C**

(*Describe, explain, or determine simple probabilities or outcomes of an experiment*)

The machine holds a total of 9 gumballs, and 4 of the gumballs are red, so the probability of a red gumball coming out next is 4 out of 9.

Incorrect choices:

A would be correct if only one gumball were red.

B is the number of red gumballs compared to the number of blue gumballs.

D is the probability that a blue gumball will be next.

36. Correct response: **A**

(*Make predictions or draw conclusions based on data*)

Emilio spent 11 hours reading and only 4 hours watching television, so he spent more time reading.

Incorrect choices:

B Emilio spent more time reading than watching television.

C Emilio watched more television at the beginning of the week than at the end.

D Emilio read the same number of hours at the beginning of the week and at the end.

37. Correct response: **C**

(*Identify and represent visual and number patterns using words, variables, tables, and graphs*)

In 2000, the population of Goodville was 100. In 2001, it was $100 + 300 = 400$; in 1992, it was $400 + 300 = 700$; and in 1993, it was $700 + 300 = 1,000$. The population grows by 300 each year.

Incorrect choices:

A An increase from 100 to 400 does not double the population.

B An increase from 100 to 400 does not triple the population.

D is the opposite of what the graph shows (an increase of 300 every year, not 100 every 3 years).

38. Correct response: **B**

(*Identify, describe, and extend numerical and geometric patterns*)

In the pattern, the arrows point first right, then left, then up, then down. The next arrow after an arrow that points right is an arrow that points left.

Incorrect choices:

A The arrow points right; this is the 1st, 5th, and 9th shape in the pattern, and the last figure shown.

C The arrow points up; this is the 3rd, 7th, and 11th shape in the pattern.

D The arrow points down; this is the 4th, 8th, and 12th shape in the pattern.

39. Correct response: **A**

(*Solve simple number sentences or equations with one variable*)

To find the missing number, use division, which is the inverse of multiplication: $12 \div 4 = 3$.

Incorrect choices:

B reflects an error in multiplying; $4 \times 4 = 16$, not 12.

C is the result of subtracting $12 - 4$ instead of dividing.

D is the result of multiplying 12×4 instead of dividing.

40. Correct response: **D**

(*Use objects, symbols, and words to model concepts of variables, expressions, equations, inequalities*)

To find the total number of crackers Debbie bought, multiply the number of crackers in each box times the number of boxes (10) she bought. Since the symbol in the expression represents the number of crackers in each box, the correct expression is:

$10 \times \boxed{}$.

Incorrect choices:

A, **B**, and **C** do not model the problem situation.

41. Correct response: **B**

(*Represent and describe mathematical relationships with lists, tables, charts, graphs, and diagrams*)

The height of the bean plant will be 3 inches greater each week. The data in answer choice B show this relationship: 5 in. + 3 in. = 8 in., and 8 in. + 3 in. = 11 in.

Incorrect choices:

A Shows the plant at the same height each week.

C Shows the plant growing 3 inches in week 1 and then 6 inches in week 2.

D Shows the height of the plant tripling each week.

42. Correct response: **C**

(*Identify reasonable solutions*)

To find the best estimate of the number of seats, round the number of rows (18) to 20 and round the number of seats per row (11) to 10. Then multiply: $20 \times 10 = 200$.

Incorrect choices:

A is the result of rounding to 20 and 10 and then adding instead of multiplying.

B is the result of rounding each number down to 10 and multiplying 10×10.

D is the result of rounding 11 to 10 and making an error in multiplication (adding a zero to 18×10).

Test ③ Answer Key

1. A	**10.** D	**19.** B	**28.** A	**37.** D
2. B	**11.** A	**20.** D	**29.** A	**38.** C
3. C	**12.** D	**21.** D	**30.** D	**39.** A
4. D	**13.** C	**22.** C	**31.** C	**40.** B
5. C	**14.** B	**23.** C	**32.** B	**41.** B
6. B	**15.** C	**24.** D	**33.** D	**42.** D
7. D	**16.** B	**25.** A	**34.** A	
8. B	**17.** C	**26.** A	**35.** C	
9. C	**18.** A	**27.** C	**36.** B	

Answer Key Explanations

1. Correct response: **A**
 (*Read and write whole numbers*)
   The number 5,098 has "five" in the thousands place, nothing in the hundreds place, "nine" in the tens place (ninety), and "eight" in the ones place: five thousand ninety-eight.

 Incorrect choices:
 B This number is 598.
 C This number is 5,908.
 D This number is 5,980.

2. Correct response: **B**
 (*Read and write whole numbers*)
   The standard form of two thousand four hundred seven will have a **2** in the thousands place (**2,**000), a 4 in the hundreds place (400), a zero in the tens place, and a 7 in the ones place: 2,407.

 Incorrect choices:
 A This number is two hundred forty-seven.
 C This number is two thousand four hundred seventy.
 D This number is two thousand forty-seven.

3. Correct response: **C**
(*Compare and order whole numbers*)
    To order these whole numbers, first consider the hundreds place, then the tens place, and then the ones place. Answer choice **C** shows the numbers in order from least to greatest.

Incorrect choices:

A has the digits in the tens place in order from least to greatest (1, 2, 4, 6).

B has the digits in the ones place in order from least to greatest (2, 3, 5, 7).

D shows the numbers in order from greatest to least.

4. Correct response: **D**
(*Use counting and skip counting, e.g., by 2's, 5's, 10's*)
    "D" is the 17th letter in the phrase.

Incorrect choices:

A represents a mistake in counting; "A" is the 15th letter.

B represents a mistake in counting; "N" is the 16th letter.

C represents a mistake in counting; "H" is the 18th letter.

5. Correct response: **C**
(*Use grouping to represent whole numbers*)
    Seven sets of 10, or 7 × 10, equals 70, and four 1's is 4: 70 + 4 = 74.

Incorrect choices:

A represents seven 1's + four 1's = 11.

B is the result of adding 7 + 10 + 4.

D represents seven 100's + four 1's.

6. Correct response: **B**
(*Identify and represent fractions*)
    This figure is divided into 6 equal sections, and 2 of the 6 sections are shaded, representing $\frac{2}{6}$.

Incorrect choices:

In **A**, the figure shows $\frac{1}{4}$ shaded.

In **C**, the figure shows $\frac{2}{5}$ shaded.

In **D**, the figure shows $\frac{3}{6}$, or $\frac{1}{2}$, shaded.

7. Correct response: **D**

(*Compare, order, convert, and find equivalent fractions*)

The figures show that $\frac{2}{4}$ is the same area as $\frac{1}{2}$, and $\frac{2}{4}$ can be reduced to $\frac{1}{2}$, so $\frac{1}{2} = \frac{2}{4}$.

Incorrect choices:

A is incorrect because $\frac{2}{4}$ is equal to, not greater than, $\frac{1}{2}$.

B is incorrect because $\frac{1}{4}$ is less than, not greater than, $\frac{1}{2}$.

C is incorrect because $\frac{1}{2}$ is greater than, not equal to, $\frac{1}{4}$.

8. Correct response: **B**

(*Identify and use place value*)

The two is in the tens place, so it represents 2×10, or 20.

Incorrect choices:

A would be correct if the 2 were in the ones place.

C would be correct if the 2 were in the hundreds place.

D would be correct if the 2 were in the thousands place.

9. Correct response: **C**

(*Read, write, and model simple decimals and relate decimal amounts of money to fractions*)

Howard has $0.50, which is equal to $\frac{50}{100}$, or $\frac{1}{2}$ dollar.

Incorrect responses:

A is incorrect because $\frac{1}{5}$ dollar would be $0.20.

B is incorrect because $\frac{1}{4}$ dollar would be $0.25.

D reflects an error in placing the decimal point; 5 dollars would be $5.00.

10. Correct response: **D**

(*Add and subtract whole numbers*)

To find the total number of people who visited the museum, add the number who visited on Saturday (5,378) plus the number who visited on Sunday (4,295): 5,378 + 4,295 = 9,673.

Incorrect choices:

A is the result of subtracting 5,378 − 4,295 instead of adding.

B reflects an error in addition (in the hundreds column).

C reflects an error in addition (in the tens column).

11. Correct response: **A**

(*Add and subtract whole numbers*)

To find the number of people in Starville who are not children, subtract the number of children (4,854) from the total population (8,643): $8,643 - 4,854 = 3,789$.

Incorrect choices:

B reflects an error in subtraction (in the hundreds column).

C reflects an error in subtraction (in the thousands column).

D is the result of adding $8,643 + 4,854$ instead of subtracting.

12. Correct response: **D**

(*Multiply and divide whole numbers*)

To find the number of seats in the movie theater, multiply the number of screens (4) by the number of seats for each screen (296): $4 \times 296 = 1,184$.

Incorrect choices:

A is the result of dividing $296 \div 4$ instead of multiplying.

B reflects an error in multiplication (not carrying the tens or the hundreds).

C reflects an error in multiplication (not carrying the hundreds).

13. Correct response: **C**

(*Add and subtract simple fractions*)

The total amount of milk Eva added was $\frac{2}{4}$ cup $+ \frac{1}{4}$ cup $= \frac{3}{4}$ cup.

Incorrect choices:

A is the amount of milk Eva added first.

B is the result of adding the numerators and the denominators of the fractions.

D is the result of adding $\frac{2}{4} + \frac{2}{4}$.

14. Correct response: **B**

(*Round whole numbers to the nearest ten or hundred*)

The number 482 rounded to the nearest ten is 480; since 2 is less than 5, it must be rounded down to 0.

Incorrect choices:

A is rounded incorrectly (downward) to the nearest hundred.

C is rounded up instead of down.

D is rounded to the nearest hundred instead of the nearest ten.

15. Correct response: **C**

   (*Estimate using whole numbers*)

   To estimate the total number of miles, round the number of miles Mrs. Liu drove on the first day to 250 and the number of miles she drove on the second day to 150. Then find the sum: $250 + 150 = 400$.

   Incorrect choices:

   A is the result of rounding to $200 + 100$.

   B is the result of rounding to $250 + 100$.

   D is the result of rounding to $250 + 200$.

16. Correct response: **B**

   (*Apply the properties of operations*)

   Using the commutative property, $3 \times 18 = 18 \times 3$.

   Incorrect choices:

   A, **C**, and **D** represent misunderstandings of the properties of operations.

17. Correct response: **C**

   (*Solve multi-step problems involving whole numbers*)

   To find the number of screws Risa uses for each chair, add the number she uses for the legs (16) plus the number she uses for the back (6): $16 + 6 = 22$. Since there are 4 chairs, she uses $4 \times 22 = 88$ screws in total.

   Incorrect choices:

   A is the result of adding the three numbers: $4 + 16 + 6$.

   B is the result of multiplying 4×16 and then adding 6.

   D is the result of multiplying 4×16 and then multiplying by 6.

18. Correct response: **A**

   (*Solve multi-step problems involving whole numbers*)

   After Amir saved 30 cookies for his family, he had $120 - 30 = 90$ cookies left. After he gave 18 to his teachers, he had $90 - 18 = 72$ cookies left. He divided the 72 cookies among 24 classmates: $72 \div 24 = 3$.

   Incorrect choices:

   B is the result of dividing 120 cookies among 24 classmates (without subtracting the numbers of cookies he saved for his family and gave to his teachers).

   C is the result of adding $120 + 30 + 18$ and then dividing the sum by 24.

   D is the result of subtracting $120 - 30 - 18 - 24$ instead of dividing in the last step.

19. Correct response: **B**

(*Tell time and find elapsed time*)

The five weeks are: July 15–21, July 22–28, July 29–August 4, August 5–11, and August 12–18.

Incorrect choices:

A is the result of counting through August 12 and omitting the last week.

C is the result of counting the number of Saturdays from July 15 through August 19.

D is the result of counting 4 weeks in July and 3 weeks in August.

20. Correct response: **D**

(*Convert units of measurement, e.g., feet and inches, meters and centimeters*)

To convert gallons to quarts, multiply the number of gallons by 4: 8 gallons $\times$ 4 quarts/gallon = 32 quarts.

Incorrect choices:

A is the result of dividing 8 ÷ 4 instead of multiplying.

B is the number of quarts in a gallon.

C is the result of adding 8 + 4 instead of multiplying.

21. Correct response: **D**

(*Select appropriate unit for measuring weight/mass, capacity, length, perimeter, and area*)

Kilometers are the most appropriate units of length for measuring distances from town to town. (One kilometer is about $\frac{6}{10}$ of a mile.)

Incorrect choices:

A, **B**, and **C** are units of length, but they are too small to measure such a large distance.

22. Correct response: **C**

(*Use rulers, thermometers, and other instruments to measure accurately*)

The arrow on the scale points halfway between the 1-lb mark and the 2-lb mark, so the block weighs $1\frac{1}{2}$ pounds.

Incorrect choices:

A is the result of reading the scale backward from 1 lb.

B is the closest whole number marked on the scale (reading downward).

D is the result of reading the scale backward from 2 lb.

23. Correct response: **C**
(*Estimate and find length, perimeter, and area*)
    The perimeter is the sum of all the sides: 6 m + 5 m + 3 m + 3 m + 3 m + 8 m = 28 meters.

Incorrect choices:

A is the result of adding the four most obvious measurements (8 m + 6 m + 5 m + 3 m).

B is the sum of five measurements, but omitting one wall of 3 m.

D is the "area" of the apartment calculated by multiplying 6 m × 8 m.

24. Correct response: **D**
(*Count money, compare units of money, and solve problems involving money*)
    Three quarters are worth 3 × 25 cents = 75 cents; 2 dimes are worth 2 × 10 cents = 20 cents; 4 pennies are worth 4 × 1 cent = 4 cents. 75 cents + 20 cents + 4 cents = 99 cents, or $0.99.

Incorrect choices:

A is the value of 1 quarters, 2 dimes, and 4 pennies.

B is the value of 3 quarters and 6 pennies (not distinguishing dimes from pennies).

C is the value of 3 quarters, 1 dime (or 2 nickels), and 4 pennies.

25. Correct response: **A**
(*Identify, classify, and describe plane figures and their attributes*)
    A pentagon has 5 sides, as in the figure shown.

Incorrect choices:

B is a trapezoid; it has 4 sides.

C is a hexagon; it has 6 sides.

D is a parallelogram; it has 4 sides.

26. Correct response: **A**
(*Identify congruent figures and lines of symmetry*)
    Answer choice **A** shows the line that divides the isosceles triangle into two congruent portions. This may be demonstrated by cutting an isosceles triangle from a piece of paper and folding along the dotted line.

Incorrect choices:

The lines in **B**, **C**, and **D** do not divide the isosceles triangle into two congruent portions and thus are not lines of symmetry.

27. Correct response: **C**

(*Identify, classify, and describe solid figures*)

A square pyramid has 5 faces: the square base and 4 triangular sides.

Incorrect choices:

A, **B**, and **D** represent misconceptions about square pyramids or a misunderstanding of the term *faces*.

28. Correct response: **A**

(*Locate points on a coordinate plane using ordered pairs*)

In an ordered pair, the first number (x) indicates the number of units across and the second number (y) indicates the number of units up or down. On the grid, point P is 2 units across and 5 units up.

Incorrect choices:

B Point Q is located at (4, 5).

C Point R is located at (5, 2).

D Point S is located at (1, 2).

29. Correct response: **A**

(*Solve problems involving volume/capacity and weight/mass*)

To find the volume of the tank, multiply its height (1 ft) times its width (1 ft) times its length (2 ft): 1 ft $\times$ 1 ft $\times$ 2 ft = 2 cubic feet.

Incorrect choices:

B is the result of adding 2 ft + 1 ft.

C is the result of adding the three measurements: 2 ft + 1 ft + 1 ft.

D is the result of adding the three measurements (2 ft + 1 ft + 1 ft) and multiplying by 2.

30. Correct response: **D**

(*Recognize, compose, and decompose 2-dimensional and 3-dimensional shapes*)

The solid object is made from two 3-dimensional shapes: a square pyramid on top of a cube.

Incorrect choices:

A lists only three 2-dimensional shapes.

B includes a 2-dimensional square instead of a cube.

C includes a 2-dimensional triangle instead of a pyramid.

31. Correct response: **C**
(*Interpret data presented in bar graphs, pictographs, tables, tallies, and charts*)
   The Bears won 6 games and lost 3, and the Lions won 5 games and lost 4; both teams won more games than they lost.

   Incorrect choices:

   A includes the Tigers, a team that won 4 games and lost 5, instead of the Lions.

   B includes the Tigers, a team that won 4 games and lost 5.

   D lists the two teams that lost more games than they won (with records of 4 and 5, and 2 and 7, respectively).

32. Correct response: **B**
(*Interpret data presented in bar graphs, pictographs, tables, tallies, and charts*)
   The bar graph in answer choice **B** is the only one that shows students owning more cats than birds; it shows that 4 students own cats and 2 own birds, and 4 > 2.

   Incorrect choices:

   A Shows that 3 students own cats and 4 students own birds; 3 < 4.

   C Shows that an equal number of students (4) own cats and birds.

   D Shows that 2 students own cats and 4 students own birds; 2 < 4.

33. Correct response: **D**
(*Use data to describe events as more, less, or equally likely*)
   Dara's data show that the light was yellow 7 times and green 23 times, so it is less likely that the light will be yellow than green.

   Incorrect choices:

   A The light was red fewer times than it was green.

   B The light was red fewer times than it was green.

   C The light was red more times than it was yellow.

34. Correct response: **A**
(*Describe, explain, or determine simple probabilities or outcomes of an experiment*)
   There are six faces on the number cube, and one of them is labeled "4," so the probability of rolling a "4" is 1 out of 6.

   Incorrect choices:

   B represents the probability of rolling a 4 with only the three faces showing in the picture.

   C represents a misunderstanding of how to find probability.

   D represents the desired number "4" over the number of possible outcomes (6).

35. Correct response: **C**

(*Describe, explain, or determine simple probabilities or outcomes of an experiment*)

The bag contains fewer green golf balls than any other color, so Miko is least likely to choose a green golf ball.

Incorrect choices:

A Miko is most likely to pick a red golf ball because there are more red balls (10) in the bag than other colors.

B There are more blue golf balls (5) in the bag than green ones (3).

D The bag contains more yellow golf balls (7) than green ones (3).

36. Correct response: **B**

(*Make predictions or draw conclusions based on data*)

Each year, there were more cloudy days than rainy days, so it is likely that there will be more cloudy days than rainy days next year.

Incorrect choices:

A There were more sunny days than cloudy days for the past three years.

C There were more sunny days than rainy days for the past three years.

D There were more cloudy days than rainy days for the past three years.

37. Correct response: **D**

(*Identify and represent visual and number patterns using words, variables, tables, and graphs*)

For every person at the fair, there are 2 raffle tickets in the jar. To find the total number of raffle tickets in the jar if *P* people are at the fair, multiply *P* by 2.

Incorrect choices:

A reflects the number of tickets for 2 people but not for any other number.

B and **C** do not fit the pattern in the table.

38. Correct response: **C**

(*Identify, describe, and extend numerical and geometric patterns*)

In the pattern, each number is 4 greater than the number before. To find the next number in the pattern, add 4 to 18: $18 + 4 = 22$.

Incorrect choices:

A is the result of adding 1 to 18 instead of adding 4.

B is the result of adding 2 to 18 instead of adding 4.

D is the result of adding 6 to 18 instead of adding 4.

39. Correct response: **A**

(*Solve simple number sentences or equations with one variable*)

The left side of the number sentence has a value of $12 - 6 = 6$. To make the right side of the number sentence equal to 6, you have to find the number that can be added to 2 to equal 6. Using subtraction, the inverse of addition,
$6 - 2 = 4$.

Incorrect choices:

B is the result of subtracting $12 - 6$ on one side of the equation.

C is the result of adding the 6 and the 2 from different sides of the equation.

D is the solution for the number sentence $12 + 6 = \_\_ + 2$.

40. Correct response: **B**

(*Use objects, symbols, and words to model concepts of variables, expressions, equations, inequalities*)

To find the total number of markers Carl has, add the number he started with (10) plus the number Sophia gave him (M): $10 + M$. You know that number is less than 15, so the inequality can be written as $10 + M < 15$.

Incorrect choices:

A and **C** do not model the problem situation.

D sets $10 + M$ as greater than 15 instead of less than 15.

41. Correct response: **B**

(*Represent and describe mathematical relationships with lists, tables, charts, graphs, and diagrams*)

Phil adds 2 stamps each month, so the number of stamps in his collection increases by 2 each month. The graph in answer choice **B** shows this relationship.

Incorrect choices:

In **A**, the graph shows the stamp collection growing by 1 stamp each month.

In **C**, the graph shows the stamp collection staying at 2 stamps each month.

In **D**, the graph shows the stamp collection growing by 2's in Month 2.

42. Correct response: **D**

(*Identify reasonable solutions*)

The total number of bagels Quincy bought is 10 times the number of bagels in each bag, so the total number of bagels has to be divisible by 10; 150 is the only answer choice divisible by 10.

Incorrect choices:

A, B, and **C** are not reasonable answers because none of them is divisible by 10.

Standardized Test Tutor: Math   Grade ③

Student Scoring Chart

Student Name _____

Teacher Name _____

Test 1	Item Numbers	No. Correct/ Total	Percent (%)
Number and Number Sense	1–9	/9	
Operations	10–18	/9	
Measurement and Geometry	19–30	/12	
Statistics and Probability	31–36	/6	
Patterns, Relations, and Algebra	37–42	/6	
Total	**1–42**	**/42**	

Test 2	Item Numbers	No. Correct/ Total	Percent (%)
Number and Number Sense	1–9	/9	
Operations	10–18	/9	
Measurement and Geometry	19–30	/12	
Statistics and Probability	31–36	/6	
Patterns, Relations, and Algebra	37–42	/6	
Total	**1–42**	**/42**	

Test 3	Item Numbers	No. Correct/ Total	Percent (%)
Number and Number Sense	1–9	/9	
Operations	10–18	/9	
Measurement and Geometry	19–30	/12	
Statistics and Probability	31–36	/6	
Patterns, Relations, and Algebra	37–42	/6	
Total	**1–42**	**/42**	

Comments/Notes:_____

_____

Standardized Test Tutor: Math   Grade

Classroom Scoring Chart

Teacher Name _____

Student Name	Test 1	Test 2	Test 3

Notes: